Advances in Intelligent and Soft Computing

64

Editor-in-Chief: J. Kacprzyk

Advances in Intelligent and Soft Computing

Editor-in-Chief

Prof. Janusz Kacprzyk
Systems Research Institute
Polish Academy of Sciences
ul. Newelska 6
01-447 Warsaw
Poland
E-mail: kacprzyk@ibspan.waw.pl

Further volumes of this series can be found on our homepage: springer.com

Ewaryst Tkacz and Adrian Kapczynski
(Eds.)

Internet – Technical Development and Applications

 Springer

Editors

Prof. Dr. Ewaryst Tkacz
Academy of Business in Dabrowa Gornicza
ul. Cieplaka 1 C
41-300 Dabrowa Gornicza
Poland
E-mail: etkacz@polsl.pl

Prof. Dr. Adrian Kapczynski
Academy of Business in
Dabrowa Gornicza
ul. Cieplaka 1 C
41-300 Dabrowa Gornicza
Poland
E-mail: prof@poczta.onet.pl

ISBN 978-3-642-05018-3   e-ISBN 978-3-642-05019-0

DOI 10.1007/978-3-642-05019-0

Advances in Intelligent and Soft Computing   ISSN 1867-5662

Library of Congress Control Number: 2009937146

Typeset & Cover Design: Scientific Publishing Services Pvt. Ltd., Chennai, India.

Printed in acid-free paper

5 4 3 2 1 0

springer.com

Preface

In the period of last two decades it is not hard to observe unusual direct progress of civilization in many fields concerning conditionality coming up from technical theories or more generally technical sciences. We experience extraordinary dynamics of the development of technological processes including different fields of daily life which concerns particularly ways of communicating. We are aspiring for disseminating of the view that the success in the concrete action is a consequence of the wisdom won over, collected and appropriately processed. They are talking straight out about the coming into existence of the information society.

In such a context the meeting of the specialists dealing with the widely understood applications of the Internet give a new dimension associated with promoting something like the new quality. Because having the information in today's world of changing attitudes and socio-economic conditions can be perceived as one of the most important advantages. It results from the universal globalization letting observe oneself of surrounding world. Thanks to the development of the Internet comprehending the distance for broadcast information packages stopped existing. Also both borders of states don't exist and finally social, economic or cultural differences as well. It isn't possible to observe something like that with reference to no other means of communication. In spite of these only a few indicated virtues, how not arousing stipulations still it is much more to be done.

At present, as it seems, implementing the universal standardization of the transfer and the processing of information is the most important issue what in the significant way influences for expanding the circle of Internet recipients of services. Because they think that the total volume of the information included in the Internet is doubling from the sequence of only three months. It means that in the sequence of the year this volume is increasing as far as sixteen times. So it seems that particularly significant is taking attempts of permanent improving parameters of Internet protocols, creating ways of establishing contact or finally reaching the agreement concerning formats of the data transfer. One should aspire to it permanent integration rather than the disintegration to progress in the context of the technological development. Hence the constant observation and the appropriate problem analysis of computer networks as well as checking the technology development on Internet with their applications is picking the great importance up.

One should assign the extraordinary weight to issues of the assurance privacies and safeties of computer services. Peculiarly it concerns the societies based on the knowledge where the information is playing the crucial role. Taking this fact into consideration it is possible without the risk to state, that more further stormy technology development influencing both Internet and its applications in the significant way is correlated with the need to ensure security of sending, processing and storing the information in computer networks. The monograph returned to hands of readers being a result of meeting specialists dealing with above mentioned issues should in the significant way contribute to the success in implementing consequences of human imagination into the social life. We believe being aware of a human weakness and an imperfection that the monograph presenting of a joint effort of the increasing numerically crowd of professionals and enthusiasts will influence the further technology development regarding Internet with constantly expanding spectrum of its applications.

Dabrowa Gornicza   Ewaryst Tkacz
November 2009   Adrian Kapczynski

Contents

List of Contributors

Malgorzata Bach
Faculty of Automatic Control,
Electronics and Computer Science,
Silesian University of Technology,
Akademicka 16,
44-100 Gliwice, Poland,
malgorzata.bach@polsl.pl

Andrzej Bialas
Research and Development
Centre EMAG,
Leopolda 31, 40-189 Katowice,
Poland,
a.bialas@emag.pl

Marcin Caban
Faculty of Automatic Control,
Electronics and Computer Science,
Silesian University of Technology,
Akademicka 16, 44-100 Gliwice,
Poland,
marcin.caban@polsl.pl

Kazimierz Choros
Institute of Informatics,
Wroclaw University of Technology,
Wybrzeze Wyspianskiego 27,
50-370 Wroclaw, Poland,
kazimierz.choros@pwr.wroc.pl

Beata Czarnacka-Chrobot
Departament of
Business Informatics,
The Warsaw School of Economics,
Al. Niepodleglosci 164,
022-554 Warszawa, Poland,
bczarn@sgh.waw.pl

Adam Czubak
Institute of Mathematics and
Computer Science,
Opole University, ul. Oleska 48,
45-052 Opole, Poland,
adam.czubak@math.uni.opole.pl

Piotr Debiec
Institute of Electronics,
Technical University of Lodz,
Wolczanska 211/215,
90-924 Lodz, Poland,
pdebiec@p.lodz.pl

Rafal Deja
Department of Computer Science,
Academy of Business in
Dabrowa Gornicza,
Cieplaka 1c,
41-300 Dabrowa Gornicza, Poland,
rdeja@wsb.edu.pl

Krzysztof Dobosz
Faculty of Automatic Control,
Electronics and Computer Science,
Silesian University of Technology,
Akademicka 16, 44-100 Gliwice,
Poland,
krzysztof.dobosz@polsl.pl

Joanna Domanska
Institute of Theoretical and
Applied Informatics,
Polish Academy of Sciences,
Baltycka 5, 44-100 Gliwice,
Poland,
joanna@iitis.gliwice.pl

Adam Domanski
Institute of Informatics,
Silesian University of Technology,
Akademicka 16, 44-100 Gliwice,
Poland,
adam.domanski@polsl.pl

Wojciech Filipowski
Faculty of Automatic Control,
Electronics and Computer Science,
Silesian University of Technology,
Akademicka 16, 44-100 Gliwice,
Poland,
wojciech.filipowski@polsl.pl

Przemyslaw Glomb
Institute of Theoretical and Applied
Informatics,
Polish Academy of Sciences,
Baltycka 5, 44-100 Gliwice, Poland,
pglomb@iitis.pl

Mirka Greskowa
Department of Information
Technologies,
College of Management/
City University of Seattle,
Panonska cesta 17,
85104 Bratislava, Slovakia,
mgreskova@vsm.sk

Andrzej Grzywak
Department of
Computer Science,
Academy of Business in
Dabrowa Gornicza,
Cieplaka 1c,
41-300 Dabrowa Gornicza,
Poland,
agrzywak@wsb.edu.pl

Katarzyna Harezlak
Faculty of Automatic Control,
Electronics, and Computer Science,
Silesian University of Technology,
Akademicka 16, 44-100 Gliwice,
Poland,
katarzyna.harezlak@polsl.pl

Daniel Jachyra
Department of Information
Systems and Applications,
University of Information Technology
and Management in Rzeszow,
Sucharskiego 2, 35-225 Rzeszow,
Poland,
djachyra@wsiz.rzeszow.pl

Adrian Kapczynski
Department of Computer Science and
Econometrics,
Silesian University of Technology,
Roosevelta 26-28, 41-800 Zabrze,
Poland,
adriank@polsl.pl

Przemyslaw Kowalski
Institute of Theoretical and
Applied Informatics,
Polish Academy of Sciences,
Baltycka 5, 44-100 Gliwice,
Poland,
przemek@iitis.gliwice.pl

Stanislaw Kozielski
Faculty of Automatic Control,
Electronics and Computer Science,

Silesian University of Technology,
Akademicka 16, 44-100 Gliwice,
Poland,
stanislaw.kozielski@polsl.pl

Vladimir Krajcik
Business School Ostrava,
Michalkovicka 1810/181,
710-00 Ostrava,
Czech Republic,
vladimir.krajcik@vsp.cz

Andrzej Materka
Institute of Electronics,
Technical University of Lodz,
Wolczanska 211/215,
90-924 Lodz, Poland,
materka@p.lodz.pl

Hanna Mazur
Institute of Informatics,
Wroclaw University of Technology,
Wybrzeze Wyspianskiego 27,
50-370 Wroclaw,
Poland,
hanna.mazur@pwr.wroc.pl

Zygmunt Mazur
Institute of Informatics,
Wroclaw University of Technology,
Wybrzeze Wyspianskiego 27,
50-370 Wroclaw, Poland,
zygmunt.mazur@pwr.wroc.pl

Teresa Mendyk-Krajewska
Institute of Informatics,
Wroclaw University of Technology,
Wybrzeze Wyspianskiego 27,
50-370 Wroclaw, Poland,
teresa.mendyk-krajewska@
   pwr.wroc.pl

Miroslaw Moroz
Department of Economics and
Organization of Enterprise,

Wroclaw University of Economics,
Komandorska 118/120,
53-345 Wroclaw, Poland,
miroslaw.moroz@ue.wroc.pl

Mateusz Nowak
Institute of Theoretical and
Applied Informatics,
Polish Academy of Sciences,
Baltycka 5, 44-100 Gliwice, Poland,
m.nowak@iitis.pl

Slawomir Nowak
Department of Computer Science,
Academy of Business in
Dabrowa Gornicza,
Cieplaka 1c, 41-300
Dabrowa Gornicza, Poland,
snowak@wsb.edu.pl

Piotr Pecka
Institute of Theoretical and
Applied Informatics,
Polish Academy of Sciences,
Baltycka 5, 44-100 Gliwice,
Poland,
p.pecka@iitis.pl

George Pilch-Kowalczyk
Department of Computer Science,
Academy of Business in
Dabrowa Gornicza,
Cieplaka 1c, 41-300
Dabrowa Gornicza,
Poland,
school@ninta.com

Andrzej Sikorski
Faculty of Electrical Engineering,
Technical University Poznan,
Piotrowo 3A,
60-965 Poznan, Poland,
andrzejs@et.put.poznan.pl

Andrzej Sobczak
Department of Business Informatics,
The Warsaw School of Economics,
Al. Niepodleglosci 164,
22-554 Warszawa, Poland,
sobczak@sgh.waw.pl

Marcin Sobota
Computer Science and Econometrics
Department,
Silesian University of Technology,
Roosevelta 26-28,
41-800 Zabrze,
Poland,
marcin.sobota@polsl.pl

Zdzislaw Sroczynski
Institute of Mathematics,
Silesian University of Technology,
Kaszubska 23,
44-100 Gliwice,
Poland,
zdzislaw.sroczynski@polsl.pl

Michal Strzelecki
Institute of Electronics,
Technical University of Lodz,
Wolczanska 211/215,
90-924 Lodz, Poland,
mstrzel@p.lodz.pl

Michal Swiderski
Faculty of Automatic Control,
Electronics and Computer Science,
Silesian University of Technology,
Akademicka 16,
44-100 Gliwice,
Poland,
michal.swiderski@polsl.pl

Aleksandra Werner
Faculty of Automatic Control,
Electronics and Computer Science,
Silesian University of Technology,
Akademicka 16,
44-100 Gliwice, Poland,
aleksandra.werner@polsl.pl

Leonard Widmer
id Quantique SA,
ch. de la Marbrerie 3,
1227 Carouge,
Switzerland,
leonard.widmer@idquantique.com

Jakub Wojtanowski
Opole University,
Institute of Mathematics and
Computer Science,
Oleska 48, 45-052 Opole,
Poland,
jakub.wojtanowski@
  math.uni.opole.pl

Miroslaw Zaborowski
Department of Computer Science,
Academy of Business in
Dabrowa Gornicza,
Cieplaka 1c,
41-300 Dabrowa Gornicza,
Poland,
mzaborowski@wsb.edu.pl

Piotr Zawadzki
Silesian University of Technology,
Institute of Electronics,
Akademicka 16, 44-100 Gliwice,
Poland,
piotr.zawadzki@polsl.pl

Joanna Zukowska
Department of Enterprise
Management,
Karol Adamiecki University of
Economics in Katowice,
1 Maja 50, 40-287 Katowice,
Poland,
joanna.zukowska@ae.katowice.pl

Part I
Internet – Technical Fundamentals and Applications

Chapter 1
How to Improve Internet?

Daniel Jachyra

Abstract. The Internet helps us and makes our lives easier. It's very useful in many aspects of our daily lives and offers as many possibilities as the number of ideas every Internet users may have. What else can be done to this helpful tool better? There are many ways to improve Internet, like: technical improvement, limited access to web content, fight against piracy, better search algorithms, catching cyber criminals, control of information overload (spam), and increase users awareness. In recent history the television was a window to the world, now it is computer with Internet connection. It's worth to take care of Internet to be more and more useful tool to better satisfy humans needs.

1.1 How Internet Works

Today's Internet is more complex than ever before. Some people use it only to send e-mail others may use it to make big international projects. Increasingly, sentence "without Internet like without hands" is gaining importance not only for big firms but for a single 'home user' as well. Everyone wants to be 'on-line' and use worldwide Internet services [1]. This global net not only connects PCs and laptops but also many other devices like: mobile phones, palmtops, robots and home appliances. To improve it and to make something better we have to know how does it work and where does it come from? First of all, I want to present a short chronology of the most important events which had influenced Internet development.

Daniel Jachyra
University of Information Technology and Management in Rzeszow,
Department of Information Systems and Applications,
2 Sucharskiego Street, 35-225 Rzeszow, Poland
e-mail: djachyra@wsiz.rzeszow.pl
http://www.wsiz.rzeszow.pl

E. Tkacz and A. Kapczynski (Eds.): Internet – Technical Development and Appl., AISC 64, pp. 3–10.
springerlink.com   © Springer-Verlag Berlin Heidelberg 2009

1.1.1 Short Chronology

- 1957 - USSR sends out the first satellite in the world, called 'Sputnik'. In response, American Ministry of Defense set up Advanced Research Projects Agency (ARPA) to regain American scientific and technological lead in the world.
- From 1960 - American government started to network together military computational centers. This infrastructure was very simple and susceptible to attacks.
- 1965 - The universities looked for various ways to exchange data and knowledge. Following their requirements scientist Ted Nelson described the syntax of hypertext.
- 1969 -Network Working Group which established the first communication protocol was organized. Arpanet was based on the first four network nodes.
- 1971 - Arpanet has now 15 nodes. Telnet and FTP protocols are developed.
- 1971 - Response to Arpanet is French project called Cyclades.
- 1972 - Ray Tomlinson creates first e-mail program.
- 1973 - based on Cyclades project results published the first TCP (Taransmission Control Protocol) protocol specification
- 1974 - First use of the word "Internet" in TCP specification. The Cyclades network is complete and ready to use.
- 1977 - TCP was divided into TCP and IP.
- 1983, January - TCP/IP protocol is replaced by NCP.
- 1984 - Domain Name System (DNS) is created.
- 1985 - First domain in the world was registered. It was nordu.net .
- 1986 - Different networks cannot establish communication to each other. As a solution Internet-Backbone NSFnet was established.
- 1989, March - Tim Berners-Lee wrote the first version of his paper "Information Management: A Proposal". It was the first introduction to World Wide Web.
- 1990 - Military Arpanet offered to public use.
- 1990, November - Tim Berners-Lee and Robert Cailliau published the concept of world wide range hypertext project.
- 1992, December - Mosaic Browser developer by NCSA.
- 1993, October - there are about 500 Web servers worldwide.
- 1994 - First time, the number of commercial Internet users is bigger than scientific users. There are 3 millions computers on Internet.
- 1995 - Work started on IPv6 protocol, to address the problem of small number of IP addresses.
- 1997 - Project called Abilene for Internet2 started.
- 1997 - There are around six millions computers on Internet.
- 1999, October - There is one million registered domains.
- 2003 - In song "The Internet is for porn" from musical "Avenue-Q" author asked for the first time about sense and meaning of Internet.
- 2007 - Amount of spam in e-mails reached 90 percent.

Today's shape of the Internet was formed by several organizations and committees. To this group belongs among others: Federal Research Internet Coordinating Committee, Internet Activities Board and the Federal Networking Council. These groups were responsible for rules and standards that make it possible to connect together different computer networks.

1.1.2 Social Aspects

Many experts describe Internet as one of the biggest advance in information transfer since discovery of printing. This event continuously influences different aspects of our everyday lives. Present Internet is millions of times more intricate than first ARPANET.

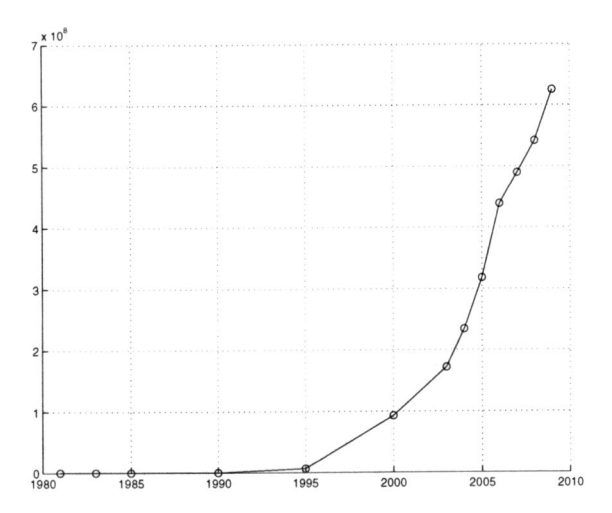

Fig. 1.1 Increase number of hosts on Internet

At the beginning of nineteen eighties e-mail nets, based on remote data transfer through telephone networks or special nets created for this purpose, were formed.

However, this technique was accessible only for universities. Internet developed and became standard of spread the information in every field of knowledge just in moment of commercial dissemination of e-mails and WWWs. Using the World Wide Web, the Internet became much more accessible and popular. Fig. 1.1 shows that the number of Internet users increases every year [2].

From the beginning of e-mail communication, the main goal was to improve commerce. Growth of bandwidth, decrease of prices and availability of link management facilitated larger dissemination of datasets. An undesirable consequence of this development was a growth of law-breaking and computer piracy.

Grows of online-journalism created a big competition for traditional mass-media. Currently we can observe change of Internet user activities from passive surfing to active author of Web 2.0 technology. Borders are replaced on Internet with thematic groups. The existing multitude of information sources requires greater responsibility and competence. This is necessary for instance in the context of political discussions when Internet is treated as a place of free expression with almost full anonymity.

Services like MySpace give possibility to create society networks. Together with growth of Internet use, in media appears information about Internet addiction. That's controversial subject among scientists. If or when extensive use of Internet has "adverse effect" or leads to abuse is still current subject of research.

Introduction of technology enabling communication by email turned out to have bad sides too. For one, it provided uncontrolled flow of information. However the problem is not only huge information flow but how to secure important data. A significant influence on this has country's law. For our own safety Internet at least in part should be monitored and controlled.

1.1.3 Internet Infrastructure

One of the funniest things about the Internet is that nobody really owns it but it doesn't mean it is not monitored and maintained. The Internet Society group supervises the policies, rules and protocols that define how to interact with the Internet. Big and small networks connect together in many different ways to form the single entity that we call Internet.

Every computer that is connected to the Internet is a part of a large network. Typical schema shows Fig. 1.2. I use a cable to connect to an Internet Service Provider (ISP). At work I am a part of LAN but I still connect to Internet using an ISP that my university has agreement with. When I connect to ISP I am a member o

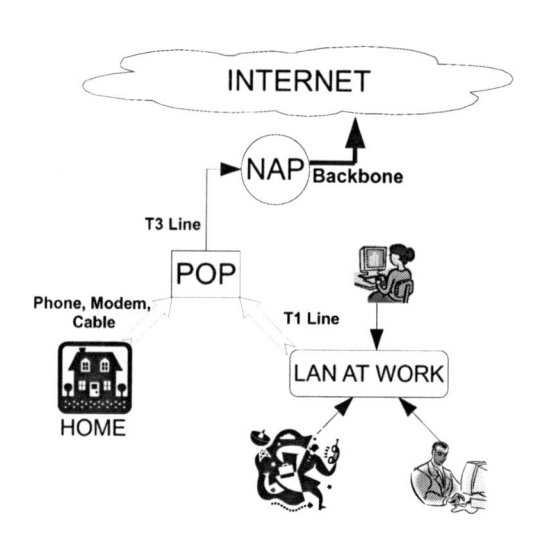

Fig. 1.2 Typical Internet connection from house and work

their network. Then ISP may connect to a larger network and became part of theirs. Therefore Internet is called network of networks.

Usually large firm has its own dedicated backbone connecting servers operated by different ISPs. This connection can provide very fast communication speeds. Service providers over multiple regions have connection points called POP (points of presents). The POP is where the local users get access to the company's network (often through dedicated line or phone number). There is no overall controlling network here. The key component of backbone is NAP (Network Access Point). It's where connections determine how traffic is routed. NAPs are the points of most Internet connection.

1.2 Problems and Their Solutions

Internet is very useful but it is not free of problems. There are as many problems as possibilities to investigate them. I want to present two of them and give some advices and propose solutions.

1.2.1 Spam Is Incredibly Annoying

The simplest spam definition is this unwanted e-mail usually of a commercial nature. It's almost always easy to recognize because of strange, illogical, full of mistakes and absurd content of e-mail. Good example of spam is e-mail with offers items at incredibly low prices, items that have no use, or services that are illegal in nature [3]. Spam is a huge problem for anyone who gets e-mail (Fig. 1.3), especially for companies and universities.

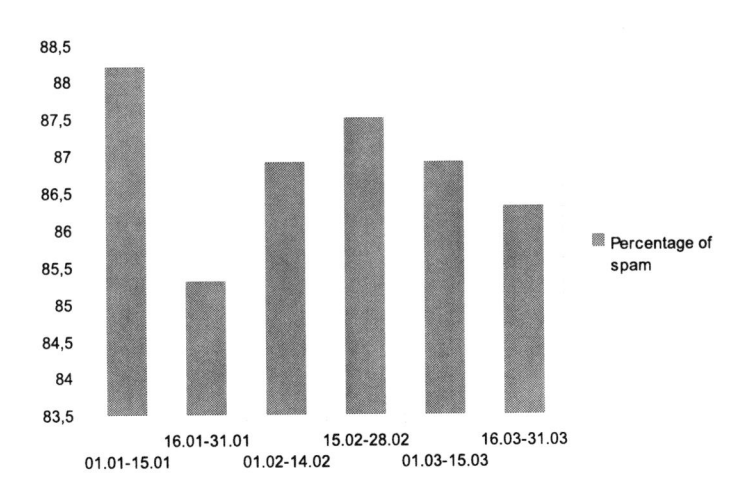

Fig. 1.3 Spam evolution: January - March 2009, source: Kaspersky Lab

Spam in mail traffic it more than 80% so it's relatively high. Spammers are now more concerned with the quantity rather than the quality of e-mails. They are trying to send as much spam as possible but pay less attention to tricks used to evade spam filters. Spamming is a business. The spammers send spam because it works. It works because although the response rate is relatively low the cost is practically nothing. The best technology that is currently available to avoid spam is filtering software. More advanced filters (heuristic filters and Bayesian filters) are based on word patterns or word frequency. Unfortunately, there are still ways to get around them. Bayesian email filters take advantage of Bayes theorem which shows suspected email by evaluating probability of a message being a spam message. Example: Suspected e-mail contains the world "warez". Most people know that this message is likely to be spam, more precisely information about portal with illegal content usually with pre-paid access. Anti-spam software, however does not "know" about this facts but it can compute probabilities that message containing a given word is spam [4]. The general formula:

$$P(A|B) = \frac{P(B|A) \cdot P(A)}{P(B|A) \cdot P(A) + P(B|C) \cdot P(C)} \, . \qquad (1.1)$$

- $P(A|B)$ - probability that a message is a spam (word "warez" is in it);
- $P(A)$ - overall probability that any given message is spam;
- $P(B|A)$ - probability that the word "warez" appears in spam messages;
- $P(C)$ - is the overall probability than any given message is no spam (is "ham")
- $P(B|C)$ - probability that the word "warez" appears in ham messages

1.2.2 There Are Necessary Better Search Engines

Internet search engines are special sites (like www.google.com) on the Web that are designed to help people find information stored on other sites. Fig. 1.4 shows how web search engines work. Without sophisticated search engines it would be practically impossible to find anything on the Web unless we know specific URL address. There are three types of search engines [5]:

- powered by crawlers (robots, spiders) - such as Google, create their listings automatically. They "crawl" the web, then people search through what they have found.
- powered by human submission - rely on humans to submit information that is subsequently indexed and catalogued. Only information that is submitted is put into the index.
- and hybrid of those two

Early search engines held an index of a few hundred thousand pages and documents, and received maybe one or two thousand inquiries each day. Today, a top search engine will index hundreds of millions of pages, and respond to tens of millions of queries per day [6].

Fig. 1.4 Web search engine

Often the same search on different search engines produce different results, why? First of all is because not all indices are going to be exactly the same. It depends on what the spiders find or what the humans submitted. Second, not every search engine uses the same algorithm to search through the indices. The algorithm is what the search engines use to determine the relevance of the information in the index to what the user is searching for.

1.3 Conclusion and Future Plans

Bill Gates in 2005 told about spam: "It will soon be a thing of the past". This anti-spam war is still in progress. The same is to search engine technology. Is still most definitely in its infancy; how it grows will very much depend upon how much information and privacy the average search engine user is willing to give up.

In both of these problems is good to apply Long Term Memory in Self-Organizing Neuronal Networks. Advanced machine learning focuses on machines reactive behavior. Recurrent neural networks (RNNs), provide general sequence processors capable of sequence learning. They use adaptive feedback links to build temporal memories. However, RNNs have limited temporal abilities. But, developed by Janusz Starzyk, new network structures called Long Term Memory (LTM) overcome the difficulties of RNNs [7]. They can efficiently solve tasks of remembering long event-driven sequences, provide contextual interpretation of text or language messages, and provide a mechanism to learn complex motor skills. These LTM structures will be used for text based learning of language categories, syntax, and semantic cognition for computer communication with humans. One of the program objectives is to develop self-organizing graph structures for specialized neural networks to model active associations in LTM that can incrementally change by acquiring new knowledge [8].

LTM based recognition will be tested against hierarchical temporal memory developed by Numenta [9] to compare training effort, recognition accuracy, playback reliability and fault tolerance using benchmark applications. Numenta developed a model for how the neocortex of the human brain might work when recognizing sequences. LTM will use a competitive winner takes all process to make input prediction. In this task the prediction mechanism will be developed and tested. Context dependent prediction will be explored. A web crawler will be used to create semantic categories based on LTM and Wikipedia text files. Subsequently, grammar rules will be extracted based on question answering. Context based text understanding will be tested on selected text messages.Self-Organizing Neuronal Networks (SONN) that capture a nature of word sequences and their relationships will be developed. SONN will link associated words and provide context information based on neuron activations. It will use algorithms for sentence decomposition and analysis of words and phrases [10] and will capture correct grammatical forms of sentences.

References

1. Handley, M.: Why the Internet only just works. BT Technology Journal 24(3) (July 2006)
2. Internet Systems Consortium, https://www.isc.org/
3. Marshall, B.: How Spam Works (25.09.2003),
   http://computer.howstuffworks.com/spam.htm (July 9, 2009)
4. Graham, P.: A Plan for Spam, http://www.paulgraham.com/spam.html
5. Levene, M.: An Introduction to Search Engines and Web Navigation. Pearson, London (2005)
6. Franklin, C.: How Internet Search Engines Work. HowStuffWorks.com (September 27, 2000),
   http://www.howstuffworks.com/search-engine.htm (July 14, 2009)
7. Starzyk, J.A., He, H.: Anticipation-Based Temporal Sequences Learning in Hierarchical Structure. IEEE Trans. on Neural Networks 18(2), 344–358 (2007)
8. Starzyk, J.A., He, H., Li, Y.: A Hierarchical Self-organizing Associative Memory for Machine Learning. In: Liu, D., Fei, S., Hou, Z.-G., Zhang, H., Sun, C. (eds.) ISNN 2007. LNCS, vol. 4491, pp. 413–423. Springer, Heidelberg (2007)
9. http://www.numenta.com/
10. Kristiansen, et al. (eds.): Cognitive Linguistics: Current Applications and Future Perspectives. Mouton de Gruyter, Berlin (2006)

Chapter 2
Performance Modeling of Selected AQM Mechanisms in TCP/IP Network

Joanna Domańska, Adam Domański, and Sławomir Nowak

Abstract. Algorithms of queue management in IP routers determine which packet should be deleted when necessary. The active queue management, recommended by IETF, enhances the efficiency of transfers and cooperate with TCP congestion window mechanism in adapting the flows intensity to the congestion in a network. Nowadays, a lot of Internet applications use UDP protocol to transport the data. In the article we analyze the impact of increased UDP traffic on Active Queue Management. The article also investigates the influence of the way packets are chosen to be dropped (end of the queue, head of the queue) on the performance, i.e. response time in case of RED.

2.1 Introduction

Algorithms of queue management in IP routers determine which packet should be deleted when necessary. The active queue management (AQM), recommended now by IETF, enhances the efficiency of transfers and cooperate with TCP congestion window mechanism in adapting the flows intensity to the congestion in the network.

Adam Domański
Institute of Informatics,
Silesian Technical University,
Akademicka 16, 44–100 Gliwice, Poland
e-mail: adam.domanski@polsl.pl

Joanna Domańska
Institute of Theoretical and Applied Informatics,
Polish Academy of Sciences,
Baltycka 5, 44–100 Gliwice, Poland
e-mail: joanna@iitis.gliwice.pl

Sławomir Nowak
Academy ff Business,
Cieplaka 1C, 41–300 Dąbrowa Górnicza, Poland
e-mail: snowak@wsb.edu.pl

E. Tkacz and A. Kapczynski (Eds.): Internet – Technical Development and Appl., AISC 64, pp. 11–20.
springerlink.com   © Springer-Verlag Berlin Heidelberg 2009

Nowadays, a lot of Internet applications use UDP protocol to transport the data. In case of UDP traffic, parameters of the queue with AQM are significantly worse. In addition, the large number of network data transmitted without flow control, makes appear a completely new, unknown phenomena in the network.

In this article we present the simulation results, simulation evaluations were carried out with the use of OMNeT++ (in version 4.0) simulation framework extended with the INET package. We added some improvements to the INET implementation: new RED algorithm (with the drop from front strategy), new sets of parameters, some new statistics and distributions and traffic scenarios.

Section 2.2 gives the basic notions on the active queue management. Section 2.3 shortly presents our simulation model. Section 2.4 discusses obtained numerical results. Some conclusions are given in section 2.5.

2.2 Active Queue Management

In *passive* queue management, packets coming to a buffer are rejected only if there is no space in the buffer to store them and the senders have no earlier warning on the danger of growing congestion. In this case all packets coming during saturation of the buffer are lost. The existing schemes may differ on the choice of packet to be deleted (end of the tail, head of the tail, random). During a saturation period all connections are affected and all react in the same way, hence they become synchronised.

To enhance the throughput and fairness of the link sharing, also to eliminate the synchronisation, the Internet Engineering Task Force (IETF) recommends *active* algorithms of buffer management. They incorporate mechanisms of preventive packet dropping when there is still place to store some packets, to advertise that the queue is growing and the danger of congestion is ahead. The probability of packet rejection is growing together with the level of congestion. The packets are dropped randomly, hence only effected users are notified and the global synchronisation of connections is avoided. A detailed discussion of the active queue management goals may be found in [11].

The RED (Random Early Detection) algorithm was proposed by IETF to enhance the transmission via IP routers. It was primarily described by Sally Floyd and Van Jacobson in [9]. Its performance is based on a drop function giving probability that a packet is rejected. The argument *avg* of this function is a weighted moving average queue length, acting as a low-pass filter and calculated at the arrival of each packet as

$$avg = (1 - w)avg + wq$$

where q is the current queue length and w is a weight determining the importance of the instantaneous queue length, typically $w \ll 1$. If w is too small, the reaction on arising congestion is too slow, if w is too large, the algorithm is too sensitive on ephemeral changes of the queue (noise). Articles [9] [19] recommend $w = 0.001$ or $w = 0.002$, and [23] shows the efficiency of $w = 0.05$ and $w = 0.07$. Article [23] analyses the influence of w on queueing time fluctuations, obviously the larger w, the higher are fluctuations. In RED drop function there are two thresholds Min_{th}

and Max_{th}. If $avg < Min_{th}$ all packets are admitted, if $Min_{th} < avg < Max_{th}$ then dropping probability p is growing linearly from 0 to p_{max}:

$$p = p_{max} \frac{avg - Min_{th}}{Max_{th} - Min_{th}}$$

and if $avg > Max_{th}$ then all packets are dropped. The value of p_{max} has also strong influence on the RED performance: if it is too large, then the overall throughput is unnecessarily choked and if it's too small the synchronisation danger arises; [19] recommends $p_{max} = 0.1$. The problem of the choice of parameters is still discussed, see e.g. [21], [22]. The mean avg may be also determined in other way, see [20] for discussion.

Despite of evident highlights, RED has also such drawbacks as low throughput, unfair bandwidth sharing, introduction of variable latency, deterioration of network stability. Therefore numerous propositions of basic algorithms improvements appear, their comparison may be found e.g. in [21].

One of our previous article [1] presented analytical (based on Markov chain) and simulation models of the RED mechanism. We assumed either Poisson or self-similar traffic. Because of the difficulty in analyzing RED mathematically [25], RED were studied in an open-loop scenario. This article evaluates the behavior of the RED queue in a simple network topology with various traffic scenarios.

2.3 Simulation Model of the RED Mechanism

The simulation evaluations were carried out with the use of OMNeT++ (in version 4.0) simulation framework extended with INET package.

The OMNeT++ is a modular, component-based simulator, with an Eclipse-based IDE and a graphical environment, mainly designed for simulation of communication networks, represented by queuing networks models and their performance evaluation. The framework is very popular in research and for academic purposes [26], [27].

The INET Framework is the communication networks simulation extension for the OMNeT++ simulation environment and contains models for several Internet protocols: UDP, TCP, SCTP, IP, IPv6, Ethernet, PPP, IEEE 802.11, MPLS, OSPF, etc. [28].

The simulated network was based on the example provided with the INET package: REDtest - to evaluate the behavior of the RED queue in a simple network topology with different traffic scenario.

The INET built-in queue algorithms are tail drop queue and RED tail drop algorithm. We add few improvements to the INET implementation: new RED algorithm (with the drop from front strategy), new sets of parameters and several new statistics, distributions and traffic scenarios.

The network topology is presented on Fig. 2.1. The link between routers r1 and r2 was the bootleneck of the network (we consider its bandwith 100kbps, 1Mbps, 10Mbps, 100 Mbps). The other links had bitrates of 100Mbps. The evaluated queue

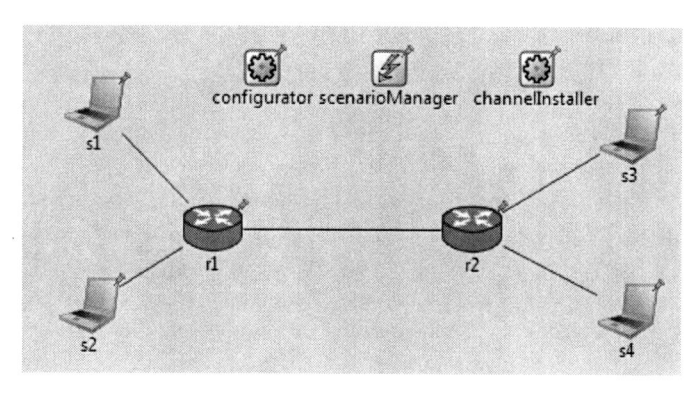

Fig. 2.1 The simulated network's topology

was the output queue for the r1 router. The simulation were perfomed for TCP connections and for a TCP operating together with UDP. The connection between hosts s1 and s3 was a TCP connection (TCP Reno). Optional we considered a UDP connection between hosts s2 and s4.

The general queue parameters were: DropQueue size = 25, RED (both cases): $w_q = 0.02$, $min_{th} = 15$, $max_{th} = 25$, $max_p = 0.03$. The simulated time was set to 200 s.

The results were obtained for the queue in router r1. This queue is the most representative in our simulated topology.

2.4 Numerical Results

In this section we present more interesting results gathered during the simulation experiments. Table 2.1, 2.2 and 2.3 present results for mixed (TCP and UDP) traffic: beetwen computers s1 and s3 - TCP Reno stream (file 2000 MB, FTP transmission), beetwen computers s2 and s4 - UDP data (1500B packet sent every 0,2 s).

Table 2.1 Simulation results: Link 100 Mbps - 100 Kbps - 100 Mbps

	Nb of packets in queue		Queue length		TCP (sender)		
	Received	Dropped (%) (RED)	Average	Std deviation	RTT Avg (s)	RTT Std Deviation	Window size (avg)
Drop	1685	52 (3%)	19,7	4,6	2,33	0,71	16971
RED tail drop	1672	79 (4,72%) (RED 20)	13,6	6,3	1,77	0,77	15564
RED front drop	1700	57 (3,35%) (RED 8)	13,5	4,7	2.16	0,61	15770

Table 2.2 Simulation results: Link 100 Mbps - 1 Mbps - 100 Mbps

	Nb of packets in queue		Queue length		TCP (sender)		
	Received	Dropped (RED)	Average	Std Deviation	RTT Avg (s)	RTT Std Deviation	Window size (avg)
Drop	16024	97	18,29	4,81	0,26	0,06	29046
RED tail drop	15792	138 (RED 110)	15,28	4,9	0,19	0.06	23736
RED front drop	16187	87 (RED 56)	15,16	2,34	0,20	0,06	24518

Table 2.3 Simulation results: Link 100 Mbps - 10 Mbps - 100 Mbps

	Nb of packets in queue		Queue length		TCP (sender)		
	Received	Dropped (RED)	Average	Std Deviation	RTT Avg (s)	RTT Std Deviation	Window size (avg)
Drop	161537	0	20,6	0,64	0,054	0,001	539399
RED tail drop	133323	199 (RED 199)	11,7	6,7	0,04	0,008	72600
RED front drop	144895	151 (RED 151)	13,6	6,9	0,039	0,009	56234

The tables don't demonstrate the results for the link (between routers s1 and s2) of 100Mbps. In this case we observe no packet loss for normal queue as well as for the queue with RED mechanism. For this case the use of AQM is not justified.

As can be concluded from the results of a very slow link (100kbps) between the routers, even a small amount of UDP transmission can not allow for the effective implementation of TCP connection.

For a large volume of traffic UDP (at 100% of link capacity) the number of dropped packets is huge and conquestion control algorithm completely stops TCP transmission. Therefore, we have reduced the speed of data generation to 1500B packet sent every 0,2 s.

The results presented in table 1 show a situation in which the AQM mechanism does not give positive results. The results do not depend on the type of queuing. Early detection mechanism works very poorly. Most packets in the buffer are discarded due to buffer overflow and only a few are rejected by RED. The distributions of the weighted moving average queue length and the queue length are presented in figures 2.3 and 2.4. Figure 2.2 shows the Round-Trip-Time parameter for the link of 100 Kbps. It can be noted, that drop from front alghorithm decreases the queue length but increases RTT.

For faster links (table 2.2 and 2.3) RED mechanism works better. In these cases the RED mechanism decreases the value of RTT (figures 2.5 and 2.8) and reduces the buffer occupancy (figures 2.7 and 2.10). The increase in the number of discarded packets in the queue caused an increase in the efficiency of the network. The improvement of network performance was evident, but not significant.

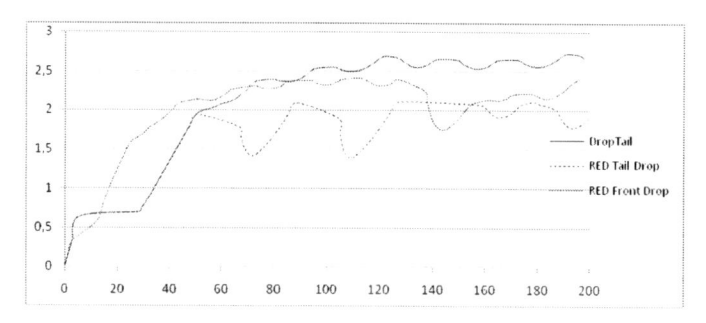

Fig. 2.2 RTT: Link 100 Kbps

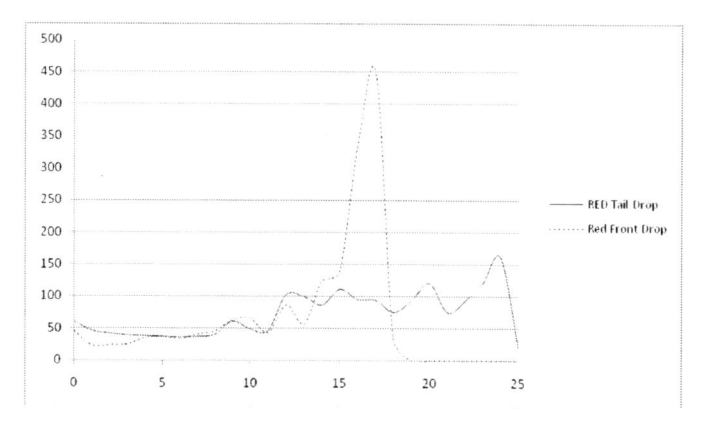

Fig. 2.3 Distribution of the weighted moving average queue length (link 100 Kbps)

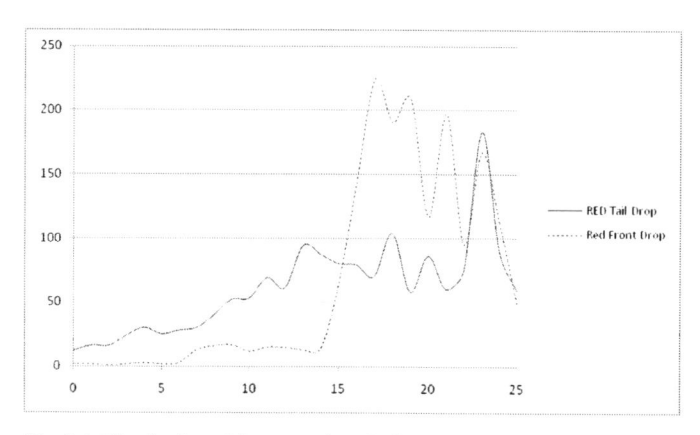

Fig. 2.4 Distribution of the queue length (link 100 Kbps)

The distribution of the weighted moving queue length (figures 2.6 and 2.9) for the RED queues is always moved slightly to the right in relation to the distribution of the RED with drop from front strategy queue. Howewer, this situation does not

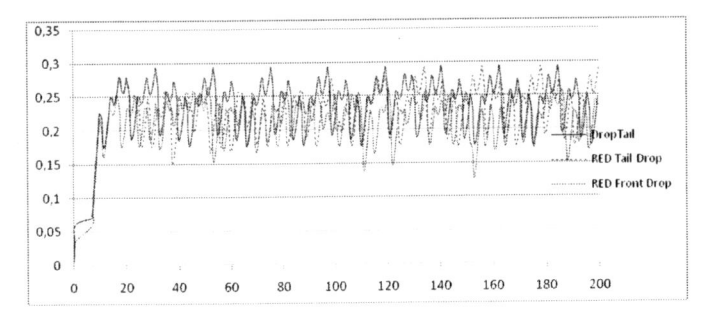

Fig. 2.5 RTT: Link 1 Mbps

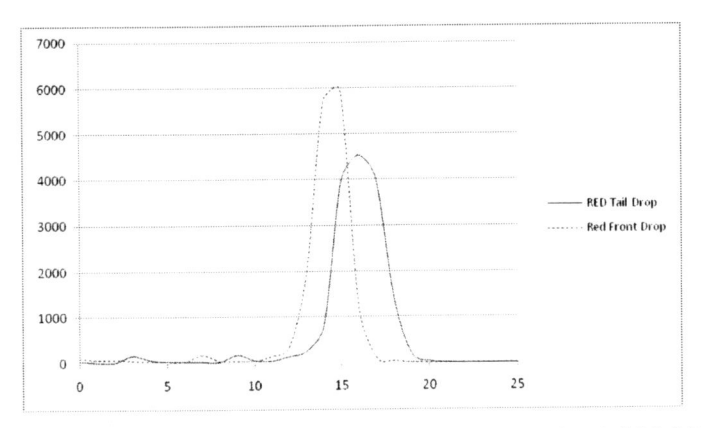

Fig. 2.6 Distribution of the weighted moving average queue length: Link 1 Mbps

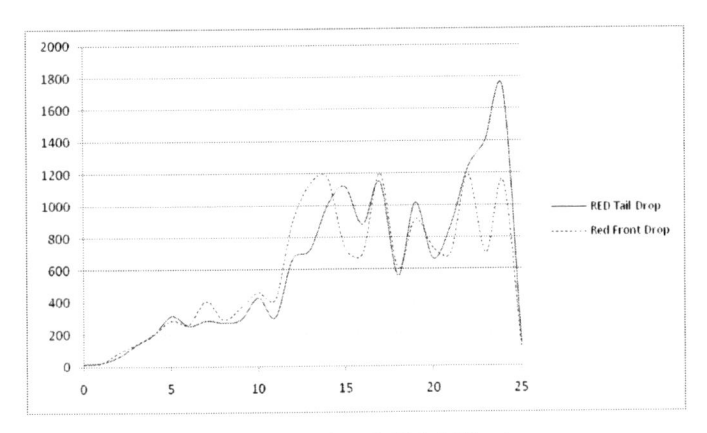

Fig. 2.7 Distribution of the queue length (link 1 Mbps)

directly bring the reduction of the RTT. However it can be clearly seen that the RTT is always smaller for the RED queues.

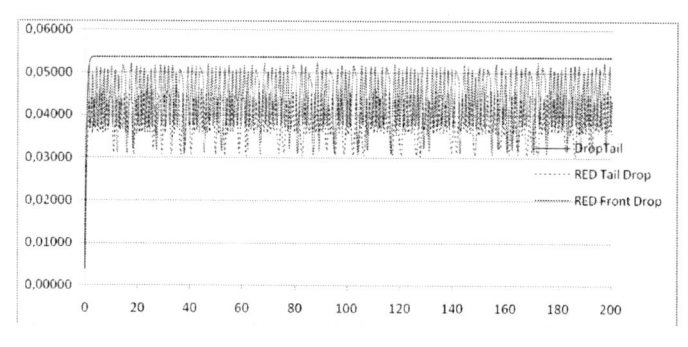

Fig. 2.8 RTT: Link 10 Mbps

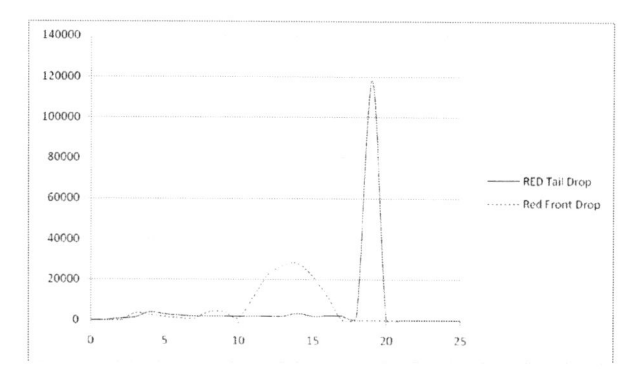

Fig. 2.9 Distribution of the weighted moving average queue length (link 10 Mbps)

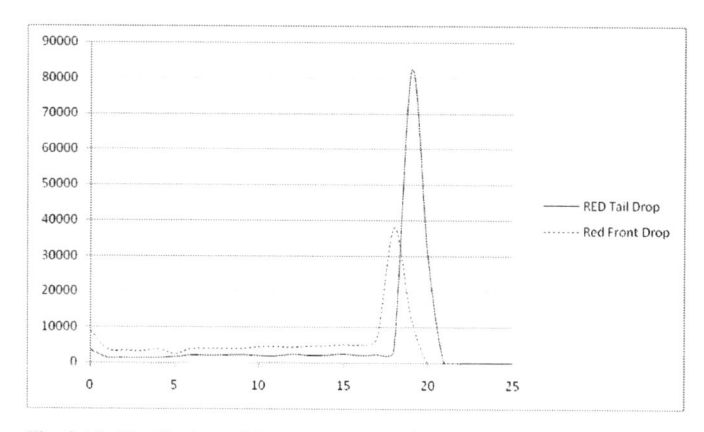

Fig. 2.10 Distribution of the queue length (link 10 Mbps)

2.5 Conclusions

In this article we present the advantages of active queue management. We also show
the influence of the way packets are chosen to be dropped (end of the queue, head

of the queue) for the behavior of the router queue for TCP, UDP and mixed (TCP, UDP) traffic. Our research was carried out in the environment of new Omnet++ (in version 4.0) simulation framework extended with the INET package.

During the tests we analyzed the following parameters of the transmission with AQM: the average length of the queue, the number of rejected packets and the RTT parameter. Classical RED algorithms are suitable only for TCP protocols with congestion algorithms. For UDP traffic the examined parameters of the transmission with AQM are significantly worse.

In addition, the article shows that AQM alghorithms and the conquestion control algorithm implemented in the TCP transmitter, in the presence of large UDP traffic, could cause long-term interruption of the TCP transmission. Today, we see an increasing percentage of UDP traffic in the Internet traffic. Applications using UDP to transfer their data (for example: P2P) may cause the serious disturbances in the network.

In this article we also investigate the influence of the way packets are chosen to be dropped (end of the queue, head of the queue) on the efficiency of the transmission. When AQM algorithms drop packets from head of the queue it is possible to notice that the number of dropped packets and the average queue length are smaller than for the case of packages dropped from the end of the queue. In our RED investigation in the *open-loop scenario* [1], dropping packets from the front of the queue also reduces the waiting time. The experiments in the TCP and UDP traffic scenario show however that smaller queue length does not necessarily reduce the RTT parameter.

Acknowledgements

This research was partially financed by Polish Ministry of Science and Higher Education project no. N517 025 31/2997.

References

1. Domańska, J., Domański, A., Czachórski, T.: The Drop-From-Front Strategy in AQM. In: Koucheryavy, Y., Harju, J., Sayenko, A. (eds.) NEW2AN 2007. LNCS, vol. 4712, pp. 61–72. Springer, Heidelberg (2007)
2. Kapadia, A., Feng, W., Campbell, R.H.: GREEN: A TCP Equation Based Approach to Active Queue Management,
   http://www.cs.dartmouth.edu/~akapadia/papers/
   UIUCDCS-R-2004-2408.pdf
3. Athuraliya, S., Li, V.H., Low, S.H., Yin, Q.: REM: Active Queue Management,
   http://netlab.caltech.edu/FAST/papers/cbef.pdf
4. Kunniyur, S.S., Srikant, R.: An Adaptive Virtual Queue (AVQ) Algorithm for Active Queue Management, http://comm.csl.uiuc.edu/srikant/Papers/avq.pdf
5. Hashem, E.: Analysis of random drop for gateway congestion control,
   http://www.worldcatlibraries.org/oclc/61689324

6. Feng, W., Kandlur, D., Saha, D., Shin, K.: Blue: A New Class of Active Queue Management Algorithms, `http://citeseer.ist.psu.edu/feng99blue.html`
7. Feng, W.H., Kandlur, D.D., Saha, D., Shin, K.G.: A Self-Configuring RED Gateway, `http://citeseer.ist.psu.edu/470052.html`
8. `http://en.wikipedia.org/`
9. Floyd, S., Jacobson, V.: Random Early Detection gateways for Congestion Avoidance, `http://www.cs.ucsd.edu/classes/wi01/cse222/papers/floyd-red-ton93.pdf`
10. Random Early Detection (RED): Algorithm, Modeling and Parameters Configuration, `http://photon.poly.edu/~jefftao/JTao_RED_report.pdf`
11. Floyd, S., Gummadi, R., Shenker, S.: Adaptive RED: An Algorithm for Increasing the Robustness of REDs Active Queue Management, `http://citeseer.ist.psu.edu/448749.html`
12. Lin, D., Morris, R.: Dynamics of Random Early Detection, `https://pdos.csail.mit.edu/~rtm/papers/fred.pdf`
13. RFC 793 - Transmission Control Protocol, `http://www.faqs.org/rfcs/rfc793.html`
14. `http://en.wikipedia.org/wiki/TCP_congestion_avoidance_algorithm`
15. Alemu, T., Jean-Marie, A.: Dynamic Configuration of RED Parameters, `http://citeseer.ist.psu.edu/728472.html`
16. Verma, R., Iyer, A., Karandikar, A.: Towards an adaptive RED algorithm for archiving dale-loss performance, `http://ieeexplore.ieee.org/xpl/freeabs_all.jsp?arnumber=1214606`
17. Yang, X., Chen, H., Lang, S.: Estimation Method of Maximum Discard Probability in RED Parameters, `http://ieeexplore.ieee.org/xpl/freeabs_all.jsp?arnumber=1712588`
18. Hong, J., Joo, C., Bahk, S.: Active queue management algorithm considering queue and load states, `http://ieeexplore.ieee.org/Xplore/login.jsp?url=/iel5/9617/30391/01401608.pdf`
19. Floyd, S.: Discussions of setting parameters (1997), `http://www.icir.org/floyd/REDparameters.txt`
20. Zheng, B., Atiquzzaman, M.: A framework to determine the optimal weight parameter of red in next generation internet routers, The University of Dayton, Department of Electrical and Computer Engineering, Tech. Rep. (2000)
21. May, M., Bonald, T., Bolot, J.: Analytic evaluation of red performance. In: IEEE Infocom 2000, Tel-Aviv, Izrael (2000)
22. Chang Feng, W., Kandlur, D., Saha, D.: Adaptive packet marking for maintaining end to end throughput in a differentiated service internet. IEEE/ACM Transactions on Networking 7(5), 685–697 (1999)
23. May, M., Diot, C., Lyles, B., Bolot, J.: Influence of active queue management parameters on aggregate traffic performance. Research Report, Institut de Recherche en Informatique et en Automatique, Tech. Rep. (2000)
24. Hassan, M., Jain, R.: High Performance TCP/IP Networking. Pearson Education Inc., London (2004)
25. Liu, C., Jain, R.: Improving explicit congestion notification with the mark-front strategy. Computer Networks (1999/2001)
26. OMNET++ homepage, `http://www.omnetpp.org/`
27. Domanska, J., Grochla, K., Nowak, S.: Symulator zdarzeń dyskretnych OMNeT++, Wyd. Wyzsza Szkola Biznesu w Dabrowie Górniczej, Dabrowa Górnicza (2009)
28. INET homepage, `http://inet.omnetpp.org/`

Chapter 3
Simulation and Analytical Evaluation of a Slotted Ring Optical Packet Switching Network

Mateusz Nowak and Piotr Pecka

Abstract. Optical Packet Switching (OPS) is a technology which thanks to its speed of operation may become popular in next-generation networks. Because optical packets are not buffered it is necessary to develop new mechanisms which would allow quality of service management in OPS networks. The paper discusses the architecture of OPS networks of the ring type with nodes enabling optical packet switching, equipped with the mechanism of Quality of Service (QoS) management. The differentiation of the quality of service was achieved by means of the proposed mechanism of priority registers with the timeout mechanism. This enables the network to be used for transporting various classes of data with varying requirements as to the quality of service. Simulation trials have been conducted and an analytical model for this type of network has been proposed.

3.1 Introduction

The technology of optical transport is used mainly in wide-range backbone networks using the SONET/SDH technology. These networks as well as the networks using other optical technologies, such as Ethernet or FDDI, belong to the class of optical-electronic networks, since packets are sent through a fibre in the optical form but they are switched in the network's nodes after being converted into the electronic form. It is likely that the next step in the development of computer networks will be fully optical networks which do not require the optical-electronic conversion in order to switch the packet. In recent years many research projects have been conducted, such as KEOPS, DAVID, ROM-EO, ECOFRAME or CARRIOCAS, which have contributed to the development of this technology which still remains in the

Mateusz Nowak · Piotr Pecka
Institute of Theoretical and Applied Informatics, Polish Academy of Sciences
Gliwice, ul. Baltycka 5
e-mail: m.nowak@iitis.pl

E. Tkacz and A. Kapczynski (Eds.): Internet – Technical Development and Appl., AISC 64, pp. 21–28.
springerlink.com
© Springer-Verlag Berlin Heidelberg 2009

experimentation stage. Because of the fact that there isn't any way of storing the light wave without its conversion into the electronic form, the basic feature of fully optical packet switches which distinguishes them from electronic devices is a complete lack of buffering packets. The ring-type network has been used in the research discussed in this paper. The ring-type architecture is easy to configure and manage, unlike the complex mesh-type networks. It is often used in city and metropolitan range networks (MAN). Its prototype was the Cambridge-Ring[1] architecture whose successors are the widely-used networks such as Token-Ring[2], FDDI[3], CRMA-II[4], ATMR[5] or RPR[6]. At the moment research is conducted in order to adapt the ring architecture to the OPS network (e.g. [[7, 8]). Suggestions presented in this paper complement this research.

3.2 Network Architecture

This paper discusses the ring-type architecture with Optical Packet Switching. The transporting unit in the network is a frame with constant capacity. Frames are transferred synchronically, in each segment of the network there is one frame, either full or empty. Network nodes are equipped with the Add/Drop mechanism which enables the packet to be removed from the frame if it has been assigned to this node or if the quality of service management mechanism has requested this, to be added to an empty frame if the node has some information to send or to transfer the frame farther in an unchanged form if the node is a transit node, intermediary in the transmission of data between nodes. All nodes in the network are equivalent to each other. The network pattern has been presented in Fig. 3.1. The architecture discussed in this paper is based on the architecture proposed in the framework of the ECOFRAME project of the French National Research Agency (ANR) and on the architecture shown in [9] and [10]. The network node is an interface between the electronic and the optical component. Data from network customers flow into the node whose structure has been shown in Fig. 3.2. Data are divided into blocks of the constant b length. Each block has a quality of service class assigned to it (the mechanism of ensuring the quality of service is described below). For each QoS class x there is a buffer which is able to store N_x blocks.

Depending on the distance between nodes and thus on the time the information spends travelling within a optical fibre, frames transporting packets in the optical network can be of varying size. Each frame is able to receive a packet of the size of p blocks. If in the node there is p waiting blocks of a given class a packet is completed and then sent with the next free frame. If in the buffer for blocks of a given class there are blocks waiting longer than a given time, the packet is also formed despite the fact that it will not be completely filled (timeout mechanism, described in [11]). The packet contains only blocks of one class which facilitates quality of service management in transit nodes.

Fig. 3.1 Ring-type optical
network

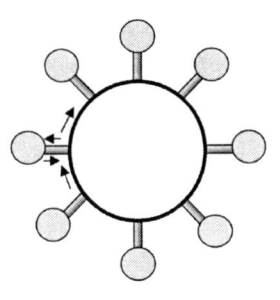

Fig. 3.2 Node of OPS ring
network with PRM

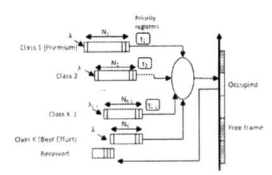

3.3 Quality of Service Mechanism

The quality of service class corresponds to the priority according to which packets enter the network. 1st class "Premium" packets are serviced with the highest priority, whereas the lowest class is "Best Effort" which includes packets sent as the last when there are no waiting packets of other packets. Premium class packets are privileged and the quality of service management mechanism should deliver them to the recipient in the shortest possible time. Best effort packets require the network to make best efforts to deliver them but they don't require the guarantee that each packet will reach the recipient as the possible mechanism of retransmitting lost packets is on the application level. The mechanism of removing "Best Effort" packets through transit nodes, based on [10] has been introduced if "Premium" packets are awaiting the access to the network in the transit node. In these circumstances, if the frame containing the "Best Effort" packet arrives at the node its content is replaced by the waiting Premium packet. Simple QoS mechanism provides that packets of lower class are sent after there are no packets of higher classes ready to send. Such a mechanism may lead to situation, that packets of class other than "Premium" are blocked by traffic of higher classes. In order to ensure the more fair differentiation of the quality of service of various class packets so-called priority registers mechanism (PRM) has been introduced. The register which exists for each class of packets except for the "Best Effort" contains a number of tokens available for the packets of a given class. The lower the number of the class, standing for a higher packet priority, the greater the initial number of tokens in the register. Each insertion of the packet of a given class from the buffer to the network decreases the number of available tokens. Highest priority packets are inserted into the network first. If all higher class tokens have been used or there is no completed packet of a higher class, packets of the next class which has a completed packet and available

tokens are inserted into the network. "Best Effort" packets are inserted only when
there are no waiting packets of a higher class. After using all the tokens in the node
they are assigned again in the initial number. Another assignment of tokens to the
initial value takes place also after T time has passed. In order to justify the introduction of the mechanism of assigning tokens again after a given time T passes let's
consider the following option: customers entering data into the network by means
of a given node don't generate information of a given q class, other than Premium
and Best Effort. After a certain time of the node's operation tokens for all classes
will be used, except for the q class. This means that it is not possible to send any
information of a class other than q or Best Effort. In order to restore the possibility
of sending packets of other classes, including the Premium class, it is necessary to
set the initial values of priority registers again after a certain time T. This mechanism should work only if within T time the values have not been set again because
of using all the tokens in the node.

3.4 Simulations

In simulation trials it has been assumed that the K number of packet servicing
classes is 4. All experiments shown in the paper were performed for the ring network
consisting of 10 nodes. Buffer sizes for each class of packets, counted in blocks of
size $b = 50$ are $N_1 = N_2 = N_3 = N_4 = 250$. Optical fibre throughput assumed is
10Gb/s. Time slot was assumed as equal to time of trip if signal between the nodes,
amounting to $1\mu s$, what gave packet size equal $p = 25$ blocks. Timeout, causing
sending of the packet despite of its incompleteness was set to 1ms, equally for all
queues. One assumed, that data incoming to network node from the client side respond to typical Internet statistics, according to [12]. Traffic for each class of client
packets is decomposed as follows: 40% of packets have of 50 bytes, 30% has size
of 500 bytes and the rest are packets of 1500 bytes of size. Packet of size greater
than 50 bytes in the moment of arrival to the node is automatically decomposed into
blocks of 50 bytes. If the number of free blocks in buffer is lesser than number of
blocks, which the client packet consists of, the packet is rejected.

Results, obtained with OMNeT++ simulator [13], comparing performance of network with and without PRM are summarized below. Probability of packet loss due
to input buffer overflow for different packet classes is shown in Table 3.4. Tables
5(a)-5(d) show average time of block stay in buffer as well as average buffer fulfilment (number of blocks).

Charts shown on Fig. 3.3 present values of probability of different queue lengths
in buffers for particular class of packets with QoS mechanism in simple version,
by very high (Fig. 3(a)), high (Fig. 3(b)), medium (Fig. 3(c)) and low (Fig. 3(d))
network load. Probability of arrival of client packet of x bytes long comes to π_x
according to Tab. 3.1.

Table 3.1 Global parameters of OPS network model

Block size	Timeslot	Optical packet size	Client packet sizes	Arrival probability of packet of given size
50 bytes	1 μs	25 blocks/1250 bytes	50,500,1500 bytes	$\pi_{50} = 0.4, \pi_{500} = 0.3, \pi_{1500} = 0.3$

Table 3.2 Simulation parameters for each class of client packets

Class	1 (Premium)	2	3	4 (Best Effort)
Buffer size N (blocks)	$N_1 = 250$	$N_2 = 250$	$N_3 = 250$	$N_4 = 250$
Timeout	$\tau_1 = 1ms$	$\tau_2 = 1ms$	$\tau_3 = 1ms$	$\tau_4 = 1ms$
Relative ratio	$R_1 = 0.25$	$R_2 = 0.25$	$R_3 = 0.25$	$R_4 = 0.25$
Absolute ratio	$A_1 = \lambda R_1$	$A_2 = \lambda R_2$	$A_3 = \lambda R_3$	$A_4 = \lambda R_4$
Initial value of priority register	4	2	1	-

Table 3.3 Parameters of simulation (both for PRM switched off and on)

Run	Very high load	High load	Medium load	Low load
Load	1.07	0.85	0.49	0.29
λ [packets/μs]	0.348	0.308	0.268	0.228

Table 3.4 Probability of packet loss for particular classes of client packets due to buffer overflow - for different network load with simple QoS and with PRM

Class/QoS	1/simple	1/PRM	2/simple	2/PRM	3/simple	3/PRM	4/simple	4/PRM
load=1.07	6.0186e-06	4.8701e-05	1.01904e-03	3.4821e-04	0.0444194	0.0125095	0.406260	0.285602
load=0.85	<1e-09	7.5464e-07	2.08993e-04	7.03543e-05	6.69894e-03	1.06854e-03	0.0900742	0.062545
load=0.49	<1e-09	<1e-09	9.67416e-06	3.7131e-06	5.23922e-05	2.84540e-05	8.09135e-4	2.57545e-04
load=0.29	<1e-09	<1e-09	<1e-09	<1e-09	<1e-09	<1e-09	1.51344e-06	6.63124e-07

3.5 Markovian Analysis

The simulation model in its simple version (without the priority register) was verified using the method of Markov processes with continuous time (CTMP). Because of a very big number of Markov chain states (resulting from a very big transition matrix) calculations were made for lesser model, however working according to the same rules described in section 2.

The ring in analytical model consisted of three nodes (R = 3, where R - number of nodes in the ring). Each of the nodes owns 2 queues (K = 2): the Priority queue (for Premium Class packets) and the Best Effort queue. The frame is able to carry a packet consisting of single block. The state vector for this model has the following components: $n1_1, n1_2, n2_1, n2_2, n3_1, n3_2, q_1, q_2, q_3, p$, where $n1_1, n2_1, n3_1$ is the number of blocks in the priority queue in the first, second and third node, $n1_2, n2_2, n3_2$ is the number of blocks in the best effort queue in the first, second and third node, the components q_1, q_2 and q_3 define the state of frames in the network and range from 0 to 2 (0 means a empty frame, 1 - a busy frame, whereas 2 - a

Table 3.5 Average time of block stay and average number of blocks in buffer for different packet classes, without and with PRM

(a) Class 1 Packets

	no PRM		with PRM	
Load	avg time	avg blocks	avg time	avg blocks
VHL	19.8838	29.6106	23.0181	35.8437
HL	20.335	27.5745	21.8163	30.418
ML	19.6203	24.4599	20.9184	20.9975
LL	20.1995	22.4317	21.1077	21.1712

(b) Class 2 Packets

	no PRM		with PRM	
Load	avg time	avg blocks	avg time	avg blocks
VHL	29.923	40.2672	26.134	37.5913
HL	26.5958	33.5174	24.129	31.7819
ML	21.6584	26.1536	20.954	22.9642
LL	20.8154	22.8697	20.935	20.1898

(c) Class 3 Packets

	no PRM		with PRM	
Load	avg time	avg blocks	avg time	avg blocks
VHL	68.7904	73.9822	67.3142	71.8654
HL	43.967	48.8495	39.1212	46.645
ML	25.0083	28.9298	23.0854	24.3657
LL	21.5983	23.4227	21.6210	21.0555

(d) Class 4 Packets

	no PRM		with PRM	
Load	avg time	avg blocks	avg time	avg blocks
VHL	248.379	136.755	242.543	132.234
HL	100.151	83.5313	97.2002	80.9314
ML	30.8378	33.6567	31.0412	33.826
LL	22.5859	24.1338	22.1843	23.8543

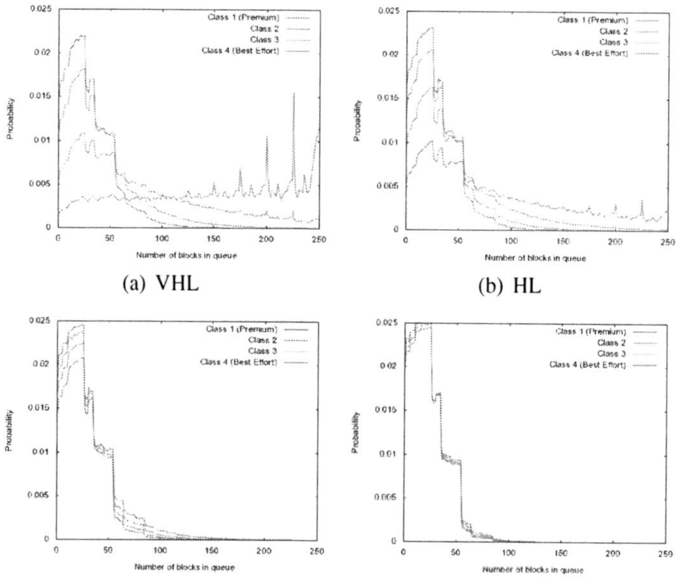

(a) VHL

(b) HL

(c) ML

(d) LL

Fig. 3.3 Probability of queue length in buffers for particular class of packets with different network load

frame transporting Best Effort packet, which can be disposed). The p value stands for the phase of Erlang's distribution which for 5 phases approximates the constant distribution very well: phases change with the intensity of $\mu = k/T$ where k is the number of phases and T is the continuous time of service, ranging from 1 to 5. If the component p is equal to 5, it means that time T has passed.

1-block packets (of the size of 50 B) arrive to the queues with the intensity of λ. The capacities of all the queues are 8 blocks. So the queues in individual nodes have the limitations of $N1_1 = N1_2 = N2_1 = N2_2 = N3_1 = N3_2 = 8$.

Possible transitions between states and transition values are shown below (for the purpose of simplification transits for the first node are shown; for the remaining nodes respective transitions are the same):

1. If $n1_1 < N1_1 \Rightarrow (n1_1 + 1, n1_2, \ldots, q1, q2, q3, p)$ with the intensity of λ – packet of the length of one block enters the Priority queue;
2. If $n1_2 < N1_2 \Rightarrow (n1_1, n1_2 + 1, n21, n22, n31, n32, q1, q2, q3, p)$ with the intensity of λ – packet of the length of one block enters the Best Effort queue;
3. If $ph < 5 \Rightarrow (n11, n12, \ldots, q1, q2, q3, p+1)$ with the intensity of μ – Erlang phase is changed;
4. If $ph = 5$ and $n1_1 > 1$ and $q_1 = 0 \Rightarrow (n1_1 - 1, n1_2, \ldots, 1, q_2, q_3, 1)$ with the intensity of μ; Premium block from priority queue is sent over the network to another node and removed from the queue;
5. If $ph = 5$ and $q_1 = 1 \Rightarrow (n1_1, n1_2, \ldots, 0, q_2, q_3, 1)$ with the intensity of $1/(R-1) * \mu$; frame containing Premium block addressed to this node arrived;
6. If $ph = 5$ and $q_1 = 2 \Rightarrow (n1_1, n1_2, \ldots, 0, q_2, q_3, 1)$ with the intensity of $1/(R-1) * \mu$; frame containing BE block addressed to this node arrived;
7. If $ph = 5$ and $n1_1 > 1$ and $q_1 = 2 \Rightarrow (n11 - 1, n12, \ldots, 1, q_2, q_3, 1)$ with the intensity of $1/(R-1) * \mu$; BE block arriving from another node is replaced by Premium block;
8. If $ph = 5$ and $n1_1 = 0$ and $n1_2 > 0$ and $q2 = 0 \Rightarrow (n1_1, n1_2 - 1, \ldots, 2, q_2, q_3, 1)$ with the intensity of μ; frame containing BE block is sent to the another node;
9. If $ph = 5 \Rightarrow (n11, n12, \ldots, q3, q1, q2, 1)$ with the intensity of μ; the frame transits to the next node (elements q_1, q_2, q_3 are shifted right $q_1 \rightarrow q_2, q_2 \rightarrow q_3, q_3 \rightarrow q_1$).

The number of states for this Markov chain is expressed by the following formula:

$$S = 5 * (3 * (N1 + 1) * (N2 + 1))^R$$

where: $N1$ - size of Priority queue, $N2$ - size of BE queue, 5 - number if Erlang phases, 3 - number of possible q_i states, R - number of nodes in a ring.

For the buffer sizes taken ($N_1 = N_2 = 8$ blocks), the number of states amounts to 71744535. The matrix of such a big dimension defines a system of linear equations. The probabilities of states are a solution of this system of equations and on their basis the remaining network parameters can be defined, such as the probabilities of loss and average queue lengths. The analytical solution makes it possible to partially verify the simulation solution - comparative results are not shown due to paper size limitations. Such a big probability vector could be determined thanks to using the modified OLYMP library, created in IITiS PAN for the purpose of big markovian model analysis [14] and adopted for work in cluster environment. The results of simulation were consistent with the results of the analytical method within an accuracy of 6 decimal places at least.

3.6 Conclusions

The Priority Registers Mechanism (PRM) of QoS management for ring-type synchronous slotted OPS networks is proposed. It was intended for managing of traffic of different QoS classes, where one wants to treat different classes of traffic fair, not allowing to block traffic of medium classes by packets of Premium class. As simulation results confirmed by Markovian analysis show, this aim is fulfilled.

All computations, which results are shown in the paper, were performed on computing cluster "Leming" in IITiS PAN Gliwice.

References

1. Needham, R.M., Herbert, A.J.: The Cambridge Distributed Computing System. Addison-Wesley, Reading (1982)
2. IEEE Std. 802.5. IEEE Standard for Token Ring (1989)
3. Ross, F.E.: Overview of FDDI: The Fiber Distributed Data Interface. IEEE JSAC 7(7) (September 1989)
4. Lemppenau, W.W., van As, H.R., Schindler, H.R.: Prototyping a 2.4 Gb/s CRMA-II Dual-Ring ATM LAN and MAN. In: Proc. 6th IEEE Wksp. Local and Metro. Area Net. (1993)
5. ISO/IECJTC1SC6 N7873. Specification of the ATMR protocol (v.2.0) (January 1993)
6. Davik, F., Yilmaz, M., Gjessing, S., Uzun, N.: IEEE 802.17 Resilient Packet Ring Tutorial
7. Mathieu, C.: Toward Packet Oadm. WDM product group, Alcatel-Lucent presentation (December 2006)
8. Chiaroni, D.: Optical packet add/drop multiplexers: Opportunities and perspectives. Alcatel-Lucent R&I, Alcatel-Lucent presentation (October 2006)
9. Eido, T., Pekergin, F., Atmaca, T.: Multiservice Optical Packet Switched networks: Modelling and performance evaluation of a Slotted Ring
10. Eido, T., Nguyen, D.T., Atmaca, T.: Packet filling optimization in multiservice slotted optical packet switching MAN networks. In: Proc. of Advanced International Conference on Telecommunications, AICT 2008, Athens (June 2008)
11. Haciomeroglu, F., Atmaca, T.: Impacts of packet filling in an Optical Packet Switching architecture. In: Advanced Industrial Conference on Telecommunications, AICT (July 2005)
12. IP packet length distribution (June 2002),
    http://www.caida.org/analysis/AIX/plen_hist/
13. OMNeT++ homepage, http://www.omnetpp.org
14. Pecka, P.: Obiektowo zorientowany wielowątkowy system do modelowania stanów nieustalonych w sieciach komputerowych za pomocą łańcuchow Markowa. PhD thesis, IITiS PAN, Gliwice (2002)

Chapter 4
A Multi-tier Path Query Evaluation Engine for Internet Information Systems

Andrzej Sikorski

Abstract. This paper provides a technique that enables embedding of path and twig query evaluation in an application server built on top of a small footprint data manager. The technique proposed offers increased flexibility, enabling seamless integration of business rules with low level evaluation facilities. This flexibility compares with that supported by stored procedures in SQL, not available for XML. Our method leverages deferred processing of a Structural Join (SJ) that we modify so as to achieve the minimum number of IO operations. The deferred structural join provides a primitive construct that can be used by our recursive composition technique, allowing local optimization of individual location steps. This recursive composition technique takes advantage of either input set low cardinality or high selectiveness of parent-child join, resulting in a performance boost.

4.1 Introduction

The conventional client-server architecture of database information systems delegates querying tasks to database servers. This common approach is not the only option. A query evaluation engine can also be implemented as an application server on top of a suitable data manager, that in addition to searching, indexing and data storage, also supports low level concurrency and transactional processing. In fact, such an architecture could offer more flexibility, as business rules can be seamlessly integrated into the low level query evaluation facilities (e.g. structural joins). This flexibility compares to that provided by stored procedures in SQL. Note that native XML databases do not yet include support for application code embedding (i.e. triggers and stored procedures).

Andrzej Sikorski
Technical University Poznań, Faculty of Electrical Engineering,
Piotrowo 3A, 60-965 Poznań, Poland
e-mail: andrzejs@et.put.poznan.pl

E. Tkacz and A. Kapczynski (Eds.): Internet – Technical Development and Appl., AISC 64, pp. 29–36.
springerlink.com

The focus of our research pertains to the optimization of Structural Join (SJ), [1] considered in the context of its interactions with a data manager. The data manager we use takes care of tasks relevant to sequential scanning, index searches and concurrency. On the physical level, data is stored in memory pages that make up a processing unit for IO. The definitive factor for our considerations here is a property of the manager: if one data object is requested, the entire page must be fetched into RAM. Our main result is a deferred operator that evaluates Structural Join in a way that decreases the number of page fetches and allows fine tuning at the level of individual location steps. This operator, along with its specialized variants, is especially useful for a web information system whose back-end component is always implemented as an application server (i.e. application code deployed on a network server) that, in conventional architecture, relies on services provided by some SQL or XML-DB server.

4.2 Related Work

The motivation for the physical organization of XML data can be found in related work [2], [3], [4]. O'Neil et al. [2] gave a description of various evaluation scenarios in MS SQL featuring proprietary ORDPATHs labeling system and analyzed effectiveness of various auxiliary indices. We use the same label-based data organization and store document nodes in a cluster index; we do not, however, rely on any particular labeling method.

Our method to tackle IO complexity is inspired by [3], that sets forth an elaborate performance model of SJ evaluation. We are doing an optimization of the measures from [3]. As far as performance is concerned, a similar result was obtained in [4] where Holistic Twig Join (HTJ) was introduced. What they did in this work was an approach consisting in processing a path/twig query in its entirety. Unlike [4], we continue with independent processing of individual location steps. In consequence, we solve two issues typical for HTJ - sub-optimality of both parent-child joins and IO operations. A rather flawed attempt to tackle the IO complexity in [5] introduces a superfluous concept of a virtual cursor that requires explicit arithmetic on labels. Unlike [5], we use labels as an abstract data type.

4.3 Deferred Evaluation of Structural Join

The deferred operator, in contrast to that given in [1], yields only one tuple at a time and does not generate an intermediate result set. It is a primitive construct that enables implementation of simple queries joining at most two input cursors. Complex path queries, that include multiple, possibly embedded, location steps, can be obtained by a recursive composition of the primitive construct (c.f. Sec. 4.5).

Fig. 4.1 gives the pseudo code for the general, ancestor-descendant deferred operator. The full-fledged implementation of a query engine must support all variants of the join operator, varying with join type and side; however the modifications are straightforward and can be easily obtained from the general variant (c.f. [1]). There are two fundamental differences with respect to the original SJ: appends are replaced with yield operations and no result tuples are constructed. The current position is represented with only one side of the join, and constructed result tuples are not needed. The yield construct (lines 08, 10) is available in C# and enables concise notation of co-routine mode processing. Execution is suspended after the *yield*, and subsequent *movenext* resumes the processing. For languages and platforms that do not support yield, standard source code transformations can be employed.

In the case of deep-left composition of operators (c.f. [6]) after a matching descendant is located, the ancestor side is no longer needed. This also eliminates the necessity of the more expensive ancestor-ordered version of SJ. Unlike the original SJ, there is also no need to iterate the ancestors stack. Much like in the case of HTJ, we assume the current path has already been determined and the matching condition for the descendant verified.

Fig. 4.1 Deferred structural join operator yielding the descendant node as the current position. The yield operation represents an equivalent of .net and LINQ *yield return* construct (c.f. [7])

```
01      operator DqSJDesc(a,d)

02      while ( eof(a) ∧ eof(d))
03        if(a<d) //labels compared
              //according to preorder
04        while ( ¬ stack.top ⊂ a) pop()
              // pop nodes not contained in
              //stack.top subtree
05        push(a); movenext(a)
06      else
07          while (¬ stack.top ⊂ d ) pop()
08          if (stack.top ⊂ d) yield d
09        movenext(d)
          end if
10      yield eof
```

4.4 Tailored Location Steps

Contrary to HTJ, the deferred operators evaluating individual joins allow a more flexible computational strategy. Keep in mind, that [4] implies restriction pertaining to HTJ optimality in the case of parent-child join. An *XPath* query can consist of multiple location steps, with varying characteristics. Thus, we introduce two optimizing options - employing FIFO (level, preorder) index for parent-child and an overloaded descendant list iterator (i.e. *movenext* with an additional hint). Tailored and general operators are compatible with each other and can be freely composed to make up path query operators.

4.4.1 Employing FIFO Index

Joining the parent node with children is much more selective than the same with descendants - a lot can be gained with FIFO order processing. Consider the join in Fig. 4.1 - the current operator position is the (a_i, d_j) tuple and the next child d_{j+k} is located far, possibly in another memory page. Standard processing based upon preorder cluster would unnecessarily scan $k - 1$ nodes coming in between two siblings.

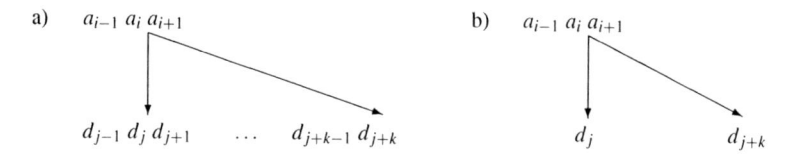

Fig. 4.2 Two siblings labeled d_j and d_{j+k} are separated by $k - 1$ descendant nodes (a). When the descendant side is processed in FIFO order, d_{j+k} is a direct successor of d_j (b)

The FIFO order makes d_{j+k} a direct successor of d_j. Due to this, we can save on a significant number of IO accesses, which are otherwise necessary to scan the subtree rooted at a_i. Now, it is sufficient to note, that the tailored operator using FIFO does not alter the output ordering. Therefore, optimization is transparent for other query components. The modifications of the general deferred operator are given in Fig. 4.3.

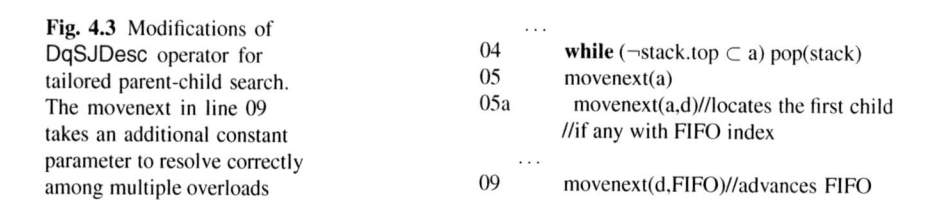

Fig. 4.3 Modifications of DqSJDesc operator for tailored parent-child search. The movenext in line 09 takes an additional constant parameter to resolve correctly among multiple overloads

```
                      . . .
04        while (¬stack.top ⊂ a) pop(stack)
05        movenext(a)
05a       movenext(a,d)//locates the first child
          //if any with FIFO index
                      . . .
09        movenext(d,FIFO)//advances FIFO
```

After the ancestor side has been advanced, the operator locates the first child (line 05), calling the appropriate variant of movenext. If there is no such node, d is positioned in such a way that it would force subsequent shift of the ancestor (in our production version we simply implemented an embedded). Observe that the second argument in descendant call (line 09) is used merely to resolve correctly among multiple overloaded operators. This variant of movenext function advances the cursor sequentially along the FIFO index - the children are located in consecutive entries.

4.4.2 Additional Hint for the Cursor Shift

For highly selective ancestor-descendant join, we may use a tailored shift function, iterating the descendant input list. This is happening with an ancestor list of much lesser cardinality than descendant. Consider Fig. 4.4 - if the gap between two subsequent ancestors exceeds the size memory page, some IO operations on the descendant side may be omitted.

The upper index of descendant node denotes the memory page ID. If $a_i << a_{i+1}$, then it is possible to skip the memory pages numbered from j to l. We are getting this by replacing sequential scanning with index positioning. For this purpose we must implement another overloaded *movenext* variant that takes the current ancestor node as a parameter. If an inspection of label values for two consecutive ancestors suggests the descendant resides on a distant page, a number of IO accesses is skipped (in Fig. 4.4 the k page would not be fetched). Moreover, most likely a_{i+1} will be co-located with its first descendant on the same page, which may result in an additional performance gain.

Fig. 4.4 A large gap between two consecutive ancestors hints that memory pages can be skipped

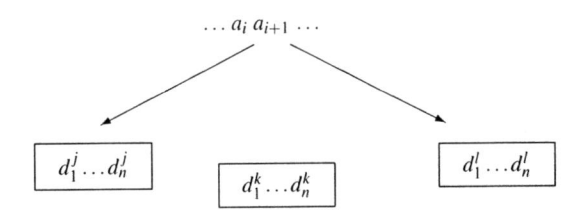

4.5 Composition of Operators

The structural join is considered to be a certain primitive operation on hierarchical data [1]. To make our result complete, we are going to show a method of operator composition, which accommodates path queries. In the conventional method (i.e. off-line) such composition is not needed because it relies on intermediate result sets. An off-line kind of query engine always processes such materialized result sets, until the final result is obtained. First, let us note that on-line operators are regular objects (i.e. instances of classes) and can be created in the usual way by invoking the constructors. The structural join constructor always takes up at least two parameters representing the input data sets. The only constraint on the input data is that it must be ordered on the join field. Consequently, only the first argument can be a result of another join; the right (i.e. descendant or child side) must be a cursor on a data table (e.g. a cursor on an auxiliary index). As data tables are stored in a cluster index the input is always guaranteed to follow preorder. Furthermore, join operator always produces an output ordered by the descendant side. Thus, both sides fulfill the constraint. Let us now consider a path query expressed in *XPath*:

```
/alpha/beta/gamma/delta/epsilon
```

Fig. 4.5 The function dist
calculates distance between
two labels; the threshold is a
function of the page size

```
04a     a1 →a
05      movenext(a)
        . . .

08      if (stack ⊃ d) yield d
08a       if dist(a1,a)<threshold
            movenext(a,d,PREORDER) else
09      movenext(d)
```

The evaluation of this query includes 5 index searches on auxiliary tag indices and 4 structural parent-child joins. If there are additional search conditions, these can be processed sequentially during the search. The translation into a sequence of SJs (and thus a composition of operators) is the following (C#):

```
IEnumerable pathQueryOp=JC(JC(JC(JC(DT("alpha"),
                                 DT("beta")),
                              DT("gamma")),
                           DT("delta")),
                        DT("epsilon"));
```

In queries containing ancestor-descendant joins, the edge JC constructor is to be replaced by its path counterpart (i.e. JD). Various kinds of constructors can be combined in the query operator, because the only type checking is performed on the *IEnumerable* interface. Thus, constructions like: JC(JD(... , ...)) are also allowed.

4.6 Experimental Evaluation

Our experimental testbed consists of two components: data manager and query evaluator. We made use of Berkeley DB, a small footprint transactional database library, supporting data persistency and indexing. We implemented the query evaluator as a native Win32 application in C++ and compiled with MS Visual Studio 2008. The testbed was deployed on a 2.2 GHz PC / 2 GB RAM and Windows Vista.

The performance evaluation was based upon a method from [1] with queries run on a large (123MB) recursive "organization" document (manager-department-employee-email). Our objective was merely confined to evaluate the deferred SJs; therefore the complete XQuery processor was not necessary. All queries were instances of a regular path pattern of the form MxDyE:

```
E       → //Employee,
M2E     → //Manager//Manager//Employee,
M1D2E   → //Manager//Department//Department//Employee.
```

An instance of XML DB Sedna [8], deployed on an identically configured PC, was used for comparison. Table 4.1 contains data about the MxD3E query execution time and result set cardinality (x ranges from 3 to 1, DQE - deferred SJ, DG - DataGuide Sedna). Fig. 4.6a shows comparison of the result set cardinality and the execution time of MxE queries (x ranges from 6 to 1) and Fig. 4.6b displays the performance of deferred SJ (DQE) with Sedna (DG).

Table 4.1 Manager-Department (MxDyE) queries performance, deferred DQE vs. DG (DataGuide -Sedna)

	DQE	DG	card.
M3D3E	2.085	4.854	558074
M2D3E	2.253	6.986	790953
M1D3E	2.542	17.312	1428389

a)   b)

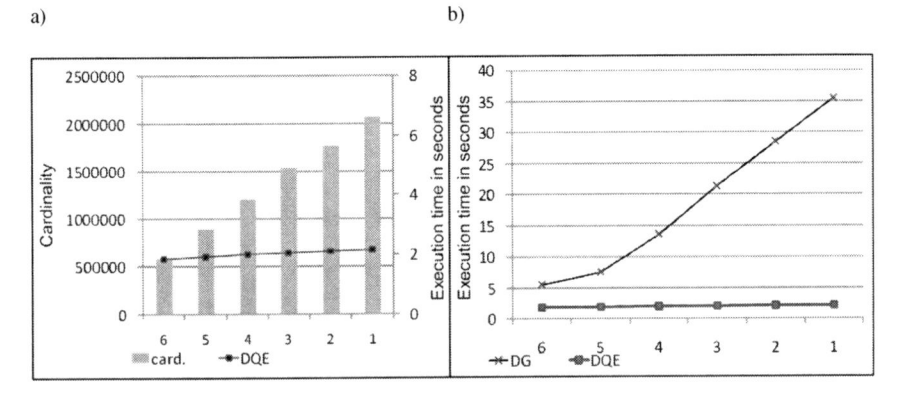

Fig. 4.6 The performance of MxE queries (DQE) vs. result set cardinality (a). The performance of MxE queries, DQE vs. DG (b)

4.7 Conclusions

The embedding of XML queries in a designated processing node (i.e. application server) is surprisingly straightforward and smoothly integrates with communication services and low level data manager processing. The usefulness of the technique given in Sec. 4.4.2 appears to be confined to the selective SJs, i.e. the path pattern has to reject some tuples from the result set. Subsequent work will show that the stepwise refinement of SJ eventually gives the PathStack [4] (and in consequence also HTJ); however, our SJ still retains its flexibility with respect to locally available optimizations. We will also show some practical programming experience in implementation of complex XML queries in the context of Apache and Berkeley DB.

References

1. Al-Khalifa, S., Jagadish, H.V., Patel, J.M., Wu, Y., Koudas, N., Srivastava, D.: Structural Joins: A Primitive for Efficient XML Query Pattern Matching. In: Proc. of the 18th International Conference on Data Engineering, pp. 141–152. IEEE Computer Society, San Jose (2002)
2. O'Neil, P.E., O'Neil, E.J., Shankar, P., Cseri, I., Schaller, G., Westbury, N.: ORDPATHs: Insert-Friendly XML Node Labels. In: SIGMOD, Paris, pp. 903–908 (2004)
3. Wu, Y., Patel, J.M., Jagadish, H.V.: Structural Join Order Selection for XML Query Optimization. In: Dayal, U., Ramamritham, K.K., Vijayaraman, T.M. (eds.) Proc. of the 19th International Conference on Data Engineering, pp. 443–454. IEEE Computer Society, Bangalore (2003)
4. Bruno, N., Koudas, N., Srivastava, D.: Holistic twig joins: optimal XML pattern matching. In: SIGMOD Conference, pp. 310–321. ACM, Madison (2002)
5. Fontoura, M., Josifovski, V., Shekita, E.J., Yang, B.: Optimizing cursor movement in holistic twig joins. In: CIKM, pp. 784–791. ACM, Bremen (2005)
6. Graefe, G.: Query evaluation techniques for large databases. ACM Comp. Surveys 25(2), 73–170 (1993)
7. Box, D., Hejlsberg, A.: LINQ:NET Language-Integrated Query, MSDN, Microsoft Corp. (2007)
8. Fomichev, A., Grinev, M., Kuznetsov, S.D.: Sedna: A Native XML DBMS. In: Wiedermann, J., Tel, G., Pokorný, J., Bieliková, M., Štuller, J. (eds.) SOFSEM 2006. LNCS, vol. 3831, pp. 272–281. Springer, Heidelberg (2006)

Chapter 5
Modeling of Internet 3D Traffic Using Hidden Markov Models

Joanna Domańska, Przemysław Głomb,
Przemysław Kowalski, and Sławomir Nowak

Abstract. The article presents the very first step of the 3D traffic modeling, which is the developing network traffic generator (equivalent to transmitting of 3D object) and methodology, based on measurements in real 3D environment (with the use of laser rangerfinder).

5.1 Introduction

Questions related to the *virtual reality* (VR) became recently more and more popular. Beside obvious examples of computer games, the VR issues are important in modern research in medicine, architecture, robotics and many others.

In practical implementation a VR system consists of a specialised software, a physical equipment and a database of *three-dimensional* (3D) objects. Usually a software realizes processing 3D object's data into the stereoscopic form (two different 2D images, representing two perspectives of the same object).

The great development of Internet in the recent years has caused that the model of work based on a single computer (with a repository of data assigned to it) obsoletes. This model was replaced by the model of the network work, in which independent computers and data repositories are connected to each other via the Internet. The problem of 3D objects distribution is related to the size of theirs models. Large size of models make the network work very difficult. The solution of this problem

Przemysław Głomb · Przemysław Kowalski · Sławomir Nowak
Institute of Theoretical and Applied Informatics,
Polish Academy of Sciences,
Baltycka 5, 44–100 Gliwice, Poland
e-mail: {pglomb,przemek,emanuel}@iitis.pl

Joanna Domańska
Academy of Business,
Cieplaka 1C, 41-300 Dąbrowa Górnicza, Poland
e-mail: jdomanska@wsb.edu.pl

E. Tkacz and A. Kapczynski (Eds.): Internet – Technical Development and Appl., AISC 64, pp. 37–44.

would be the progressive representation in conjunction with a dynamic selection of the level of the model details.

This article describes the problem of modeling of the 3D objects transmission through the Internet, including the selection of the object representation quality depending on the object distance from the virtual observer.

The whole research will consist of analytical modeling of 3D data traffic, simulation evaluation, developing of appropriate software and network protocols for exchanging 3D data in different strategies of transmission.

The model of transmission 3D objects through the Internet will be reliable - if we create a reliable model of the traffic source, which can represent the variable progression level of the 3D-meshes transmission. For the purposes of this article the choice of the progression level is based on the analysis of the spatial distribution of objects.

To create a traffic model *Hidden Markov Model* (HMM) was used. Recently, the interest in using HMM as a tool for modeling various types of network traffic is growing. However - according to the best knowledge of the authors - HMM models were not used for modeling the traffic of the 3D mesh transmissions, as it is a relatively new problem.

The article presents the very first step, which is the developing network traffic generator (equivalent to transmitting of 3D object) and methodology, based on measurements in real 3D environment (with the use of laser rangerfinder). As a result of the measurements we have a set of points, according to the distribution of physical objects in the scene. We assume that the amount of *progressive meshes* (PM [1]) data describing 3D objects is proportional to data obtained from measurements.

Section 5.2 briefly describes progressive representation of 3D objects which allows us to optimize the transmission of 3D models through the network. Section 5.3 contains a description of measurements in real 3D environment (with the use of laser rangerfinder). Section 5.4 provides a brief introduction to the HMM, and a description of its use in the creation of the traffic generator representing the variable progression levels of 3D-meshes transmission. Section 5.5 summarizes the results and presents the prospects for further work.

5.2 3D Object Representation – Progressive Meshes

There are many ways to represent 3-dimensional graphical objects, but 3D objects are usually represented as a set of meshes [2], which are a polygonal approximation of physical objects. The popular representation of meshes are triangle mesh where each face is build of 3 vertices (usually with a normal vector), and each vertex is a point of (x, y, z) coordinates.

The mesh geometry can be denoted by set (K, V) [3], where K is a simplicial complex specifying the connectivity of a mesh simplices (the adjacency of the vertices, edges, and faces), and $V = v_1, \ldots, v_m$ is a set of vertex positions defining the shape of the mesh in R^3.

Till now complex VR systems were usually local systems, meaning a 3D object's database and a visualization system were located on the same computer or within the range of local network. The size of 3D data, expressed in bytes is usually large and because of it the transmission of meshes via communication networks is a significant problem, especially taking into account real time 3D scene exploration. As examples one can mention virtual surgery or virtual museums. In such cases 3D data (meshes) are send through the network (usually based on TCP/IP protocol).

Distributed real-time systems like *Massively Multiplayer Online Game* (MMOG) are common, but network transmissions of 3D objects are limited.

In our research we focus on progressive meshes (PM) [1]. PM is a scheme for storing and transmitting triangle meshes. This representation addresses some practical problems in graphics: *level-of-detail* (LOD) approximations, progressive transmission, and mesh compression. The representation is efficient, lossless and continuous-resolution. The PM representation are useful when consider problems of large, distributed virtual reality systems, in which the distance between a source of 3D data (server) and receiver (client) is significant.

The PM is based on an arbitrary mesh M forming, stored as coarser mesh M^0 together with a sequence of n detail records, indicating how to incrementally refine M^0 back into the original mesh $M = M^n$. Each of this records store the information about a vertex split, an elementary mesh transformation adding an additional vertex to the mesh. Thus the PM representation of M defines a sequence of meshes $M^0, M^1, \ldots, M^n$ of increasing accuracy, from which LOD approximation of any desired complexity can be efficiently retrieved. Figure 5.1 presents an example of four meshes with varying degrees of complexity.

In such systems the LOD property is important, because it is easier to display the more detailed objects according to the position of a observer. Distant or hidden

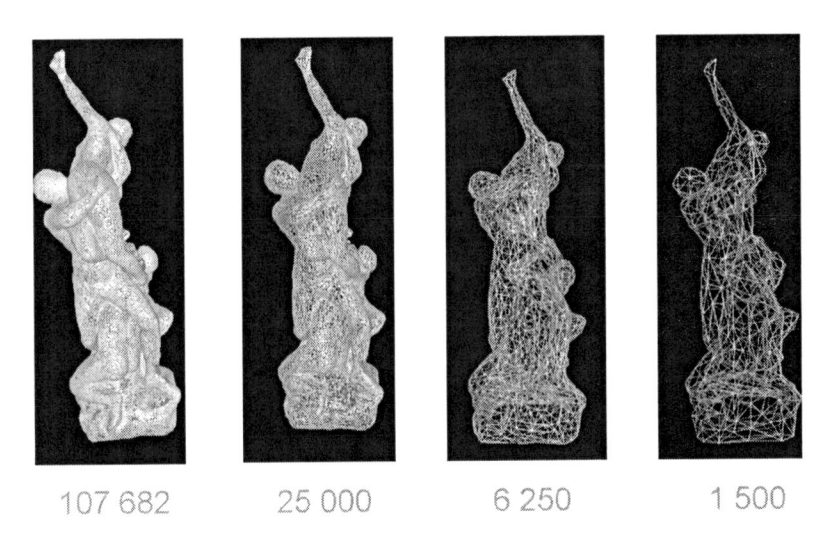

107 682   25 000   6 250   1 500

Fig. 5.1 Progressive representation of an object: a figure of Sabines from Museum in Gliwice [3]

objects don't need to be displayed with many details. While getting closer to the observer, the structures of objects are enhanced based on subsequent levels-of-details.

Recently a new model for 3D graphics compression was adopted by MPEG (as part 25) of the MPEG-4 standard [4]. The PM is now a transparent layer in the 3D graphics processing chain, easy to integrate in a end-user applications. Features such as progressive decoding or streaming are preserved.

The standard is a chance to unify the scientific effort on VR. The goal of authors is to develop a new, effective VR system, respecting problems of distributed, large TCP/IP networks. The basic assumption is that the data traffic related to the movement of an observer in virtual scene is always view-dependent. At the beginning of the research we assume that the complexity of specific objects depends only on the distance between object and the observer but not on the observer's position.

5.3 Data Acquisition

We use Pioneer P3AT robot with ARNL software [5] for preparing a map of our laboratory. The map was built in two stages: acquisition (robot is controlled by human), and map building (using software Mapper). During the data acquisition we obtain positions of the nearest objects from SICK (laser rangefinder mounted on the robot) – 177 positions (x, y) for each stage of acquisition (see Fig. 5.2). The positions are measured on the SICK altitude. The positions are registered (set in one coordination system using odometry data). The registered data creates 2D map of the laboratory. We used the data from acquisition stage of the map building, as an example of real robot navigation in an indoor environment.

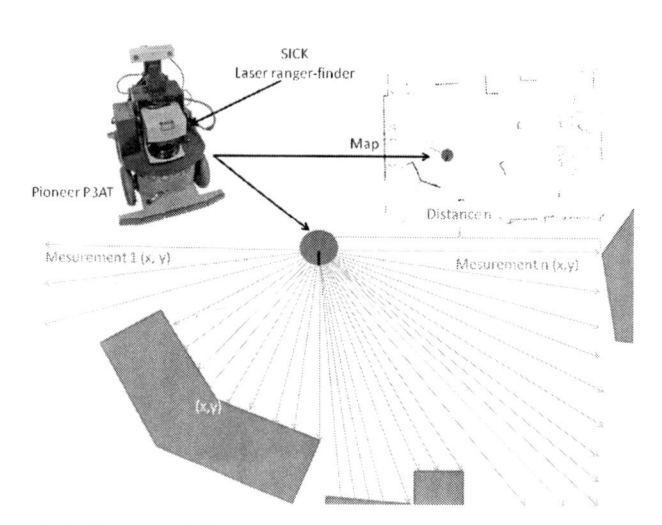

Fig. 5.2 Data acquisition

Figures 5.3 and 5.4 show histograms of distances obtained for the two complete series of acquisition. Based on the obtained data - 11 areas of the thresholds were set up, which correspond to the levels of progressive meshes details. The distance thresholds can be associated with the number of triangles forming the model [3]. This number is related to the memory complexity of transmitted models.

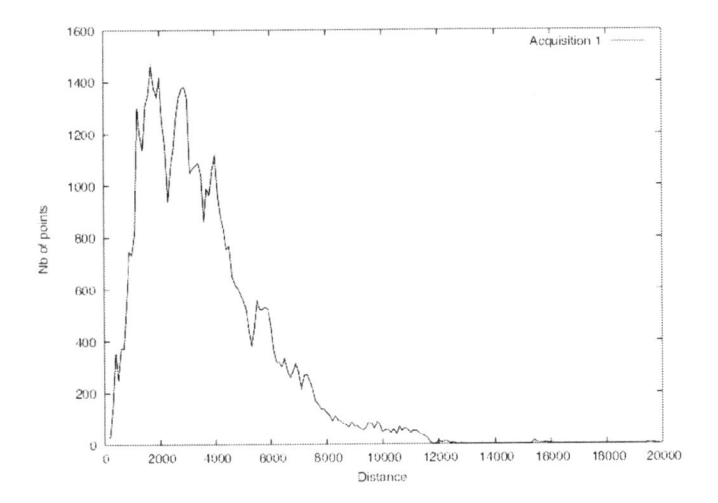

Fig. 5.3 Histogram of distances obtained for the first acquisition

In the next section we try to explain how to model the obtained sequence of views assigned to the points of robot trajectories using Hidden Markov Model.

5.4 Hidden Markov Model

Hidden Markov Model (HMM) may be viewed as a probabilistic function of a (hidden) Markov chain [6]. This Markov chain is composed of two variables:

- the hidden-state variable, whose temporal evolution follows a Markov-chain behavior ($x_n \in \{s_1,\ldots,s_N\}$ represent the (hidden) state at discrete time n with N being the number of states)
- the observe variable, that stochastically depends on the hidden state ($y_n \in \{o_1,\ldots,o_M\}$ represent the observable at discrete time n with M being the number of observable)

An HMM is characterized by the set of parameters:

$$\lambda = \{\mathbf{u}, \mathbf{A}, \mathbf{B}\}$$

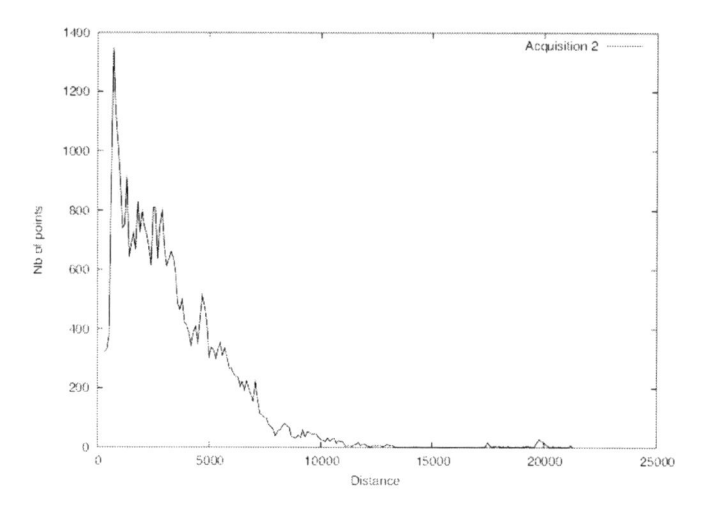

Fig. 5.4 Histogram of distances obtained for the second acquisition

where:

- **u** is the initial state distribution, where $u_i = Pr(x_1 = s_i)$
- **A** is the $N \times N$ state transition matrix, where $A_{i,j} = Pr(x_n = s_j | x_{n-1} = s_i)$
- **B** is the $N \times M$ observable generation matrix, where $B_{i,j} = Pr(y_n = o_j | x_n = s_i)$

Given a sequence of observable variables $y = (y_1, y_2, \ldots, y_L)$ referred to as the *training sequence*, we want to find the set of parameters such that the likelihood of the model $L(y; \lambda) = Pr(y | \lambda)$ is maximum. We solved it via the Baum-Welch algorithm, a special case of the Expectation-Maximization algorithm [7], that iteratively updates the parameters in order to find a local maximum point of the parameter set.

We apply a scheme using *Vector Quantization* (VQ) to translate the obtained sequence of views (more precisely: the sequence of distance histograms with 11 thresholds - see section 5.3) into a sequence of symbols, and training a HMM for this sequence. The quantization algorithm used is Linde-Buzo-Gray (LBG) algorithm of VQ [8]. Vector Quantization is a clustering technique commonly used in compression, image recognition and stream encoding [9]. It is the general approach to map a space of vector valued data to a finite set of distinct symbols, in a way to minimize distortion associated with this mapping.

We consider an HMM in which the state and the observable variables are discrete. The components of **y** represent the vector of vq-symbols components, A little portion of the sequences was used as the training sequence to learn model parameters. Performance of trained model are tested on the remaining portions of the sequences.

The obtained set of parameters is:

$$\mathbf{u} = \left[\, 0.0055 \; 0.0916 \; 0.0119 \; 0.0504 \; 0.0173 \; 0.1691 \; 0.0647 \; 0.0806 \; 0.1662 \; 0.0036 \; 0.2250 \; 0.0126 \; 0.0748 \; 0.0194 \; 0.0064 \,\right]$$

$$\mathbf{A} = \left[\begin{array}{cccccccccccccccc}
0.0117 & 0.0189 & 0.0908 & 0.0624 & 0.0815 & 0.0746 & 0.0687 & 0.1153 & 0.1362 & 0.1002 & 0.0038 & 0.0638 & 0.0175 & 0.0998 & 0.0540 \\
0.0316 & 0.0526 & 0.0786 & 0.0257 & 0.0084 & 0.1133 & 0.0074 & 0.0361 & 0.0751 & 0.0702 & 0.0768 & 0.1079 & 0.1166 & 0.1133 & 0.0857 \\
0.0213 & 0.1015 & 0.0601 & 0.0804 & 0.0818 & 0.0178 & 0.0748 & 0.0234 & 0.1169 & 0.0258 & 0.0142 & 0.0495 & 0.1138 & 0.1235 & 0.0944 \\
0.0168 & 0.0535 & 0.0776 & 0.0388 & 0.0700 & 0.0865 & 0.0569 & 0.0789 & 0.0872 & 0.0765 & 0.0739 & 0.1100 & 0.1015 & 0.0428 & 0.0282 \\
0.0113 & 0.0559 & 0.1037 & 0.0879 & 0.1622 & 0.0360 & 0.0288 & 0.0127 & 0.1067 & 0.0258 & 0.0366 & 0.1201 & 0.0864 & 0.0120 & 0.1132 \\
0.1074 & 0.0158 & 0.0299 & 0.0909 & 0.0911 & 0.0916 & 0.0614 & 0.0179 & 0.1115 & 0.0673 & 0.0228 & 0.0865 & 0.0877 & 0.0842 & 0.0332 \\
0.1096 & 0.0529 & 0.0151 & 0.0363 & 0.0909 & 0.0627 & 0.1387 & 0.0029 & 0.0810 & 0.0824 & 0.0798 & 0.1427 & 0.0275 & 0.0050 & 0.0718 \\
0.0962 & 0.0807 & 0.0828 & 0.1175 & 0.0023 & 0.1198 & 0.0304 & 0.0115 & 0.1067 & 0.0386 & 0.0607 & 0.0244 & 0.0045 & 0.1240 & 0.0991 \\
0.0309 & 0.0571 & 0.0626 & 0.0808 & 0.0535 & 0.0742 & 0.0700 & 0.0382 & 0.1050 & 0.1242 & 0.0327 & 0.0969 & 0.0322 & 0.0667 & 0.0742 \\
0.0160 & 0.0524 & 0.1446 & 0.0490 & 0.0710 & 0.0810 & 0.1073 & 0.0173 & 0.0053 & 0.1076 & 0.0542 & 0.0337 & 0.1115 & 0.0730 & 0.0751 \\
0.0752 & 0.0676 & 0.0739 & 0.0617 & 0.1114 & 0.0060 & 0.0941 & 0.0573 & 0.1258 & 0.0411 & 0.0919 & 0.0138 & 0.1266 & 0.0495 & 0.0033 \\
0.0062 & 0.0356 & 0.0690 & 0.1233 & 0.1196 & 0.0799 & 0.0559 & 0.0029 & 0.0952 & 0.0722 & 0.0230 & 0.0243 & 0.0964 & 0.1263 & 0.0694 \\
0.0005 & 0.1277 & 0.0650 & 0.0847 & 0.0512 & 0.0470 & 0.0609 & 0.0412 & 0.1201 & 0.0711 & 0.0897 & 0.1031 & 0.0201 & 0.0438 & 0.0733 \\
0.0067 & 0.0671 & 0.0144 & 0.1212 & 0.0929 & 0.0997 & 0.0852 & 0.0106 & 0.1098 & 0.0414 & 0.0241 & 0.0821 & 0.1155 & 0.0436 & 0.0850 \\
0.0416 & 0.1271 & 0.0234 & 0.0255 & 0.0041 & 0.1082 & 0.1096 & 0.0819 & 0.0033 & 0.1340 & 0.0541 & 0.0197 & 0.1235 & 0.0916 & 0.0516
\end{array}\right]$$

$$\mathbf{B} = \left[\begin{array}{cccccccccc}
0.2301 & 0.2010 & 0.0845 & 0.1620 & 0.0700 & 0.0070 & 0.0981 & 0.0285 & 0.1137 & 0.0047 \\
0.2170 & 0.1044 & 0.2275 & 0.1635 & 0.0085 & 0.0711 & 0.0641 & 0.1034 & 0.0352 & 0.0047 \\
0.3166 & 0.0736 & 0.0124 & 0.0912 & 0.1392 & 0.1210 & 0.1577 & 0.0205 & 0.0625 & 0.0047 \\
0.1043 & 0.2318 & 0.0906 & 0.0976 & 0.1566 & 0.0715 & 0.0553 & 0.0957 & 0.0915 & 0.0047 \\
0.3378 & 0.0502 & 0.1133 & 0.0286 & 0.2038 & 0.0686 & 0.0574 & 0.0602 & 0.0748 & 0.0047 \\
0.1641 & 0.1566 & 0.1275 & 0.1329 & 0.0729 & 0.0311 & 0.0579 & 0.1577 & 0.0941 & 0.0047 \\
0.3798 & 0.1196 & 0.1589 & 0.0067 & 0.0440 & 0.0314 & 0.0891 & 0.0915 & 0.0738 & 0.0047 \\
0.0462 & 0.2628 & 0.0674 & 0.1158 & 0.0233 & 0.0714 & 0.1337 & 0.1663 & 0.1079 & 0.0047 \\
0.2571 & 0.0080 & 0.0569 & 0.1858 & 0.2286 & 0.0394 & 0.0242 & 0.1618 & 0.0330 & 0.0047 \\
0.1114 & 0.0181 & 0.2196 & 0.1562 & 0.2928 & 0.0558 & 0.0752 & 0.0122 & 0.0536 & 0.0047 \\
0.1190 & 0.1054 & 0.0088 & 0.0264 & 0.1564 & 0.0370 & 0.1901 & 0.2817 & 0.0701 & 0.0047 \\
0.1813 & 0.1607 & 0.2292 & 0.0780 & 0.1496 & 0.0478 & 0.0486 & 0.0279 & 0.0716 & 0.0047 \\
0.1684 & 0.0686 & 0.1498 & 0.1170 & 0.1836 & 0.0460 & 0.0821 & 0.0935 & 0.0860 & 0.0047 \\
0.1768 & 0.1577 & 0.1746 & 0.0376 & 0.2080 & 0.0574 & 0.0620 & 0.0434 & 0.0773 & 0.0047 \\
0.2554 & 0.0482 & 0.0733 & 0.0832 & 0.1775 & 0.0801 & 0.1267 & 0.0519 & 0.0985 & 0.0047
\end{array}\right]$$

We hope that the HMM trained with more experimental data will to capture the Internet 3D traffic behavior and could be used as the 3D traffic source model.

5.5 Conclusions

In our work we assume that it is possible to develop of a new software for effective exchanging 3D data in different strategies of transmission. The article presents the first step into the research work on developing and evaluating a new VR systems, respecting the problems of network traffic and TCP/IP specificity, based on the new MPEG standard.

The parameters of the traffic generator, representing the variable progression levels of 3D-meshes transmission with the use of HMM is introduced. The generator is based on measurements with the use of laser rangerfinder. The PM related data are proportional to data obtained from that measurements.

Future works will be based on presented methodology of analytical modeling of traffic and further measurements in real environment, with the use of different type of sensors. The results will be necessary in order to create the simulation model. Based on the results we will implement the new software and evaluate its properties.

Acknowledgements

This research was partially financed by Polish Ministry of Science and Higher Education project no. N517 025 31/2997.

References

1. Hoppe, H.: Progressive meshes. ComputerGraphics (SIGGRAPH 1996 Proceedings), 99–108 (1996)
2. Nielsen, F.: Visual computing: Geometry, graphics and vision. Charles River Media (2005)
3. Skabek, K., Zabik, L.: Visual computing: Geometry, graphics and vision. Communications in Computer and Information Science (2009)
4. Jovanova, B., Preda, M., Preteux, F.: Mpeg-4 part 25: A graphics compression framework for xml-based scene graph formats. Image Communication Archive 24, 101–114 (2009)
5. Sonarnl with mobileeyes(tm). MobileRobotics Inc., Installation & Operations Manual (2006)
6. Rabiner, L.: A tutorial on hidden markov models and selected applications in speech recognition. Proceedings of the IEEE 77(2), 257–285 (1989)
7. Bilmes, J.: A gentle tutorial od the em algorithm and its application to parameter estimation for gaussian mixture and hidden markov models. Technical report, University of Berkeley (1998)
8. Linde, Y., Buzo, A., Gray, R.: An algorithm for vector quantizer design. IEEE Transactions on Communication COM-28, 84–95 (1980)
9. Romaszewski, M., Głomb, P.: 3d mesh approximation using vector quantization. Advances in Soft Computing 57, 71–78 (2009)

Chapter 6
Usability Testing of the Information Search Agent: Interface Design Issues

Mirka Greskova

Abstract. Effectiveness of information technology use is one of the key issues influencing overall organization performance. Interface design of applications needs to support users in every day work activities. We propose a general methodology for testing the quality of software applications interfaces. Methods were verified in the research aimed at the usability testing of the information search agent for advanced retrieval of internet information (Copernic Agent Professional). The main methods used in our research were think-aloud protocols, but also interviews and observations. Paper characterizes the key findings together with recommendations for application performance improvements.

6.1 Introduction

Vannevar Bush published an article *As we may think* [1] in which he described the idea of *Memex* more than sixty years ago. It was characterized as a mechanized personal directory and library storing all the necessary information. Memex was designed to broaden human memory. Its basic function was to associate related information. Probably nobody could have ever imagined that Memex will become reality. We use information from recent Memex (Internet) on every day basis.

Internet provide information access but also stores huge amount of data and information. To search, analyze and summarize relevant information is not only human,

Mirka Greskova
College of Management/City University of Seattle,
Department of Information Technologies, Panonska cesta 17,
85104 Bratislava, Slovakia
e-mail: mgreskova@vsm.sk
http://www.vsm.sk/en

E. Tkacz and A. Kapczynski (Eds.): Internet – Technical Development and Appl., AISC 64, pp. 45–50.
springerlink.com

but ever since the development of the first information retrieval systems, also machine task. Archie, which is considered as the first search engine, was developed almost thirty years ago. This tool enabled to find FTP files according to their name. It supported crawling robot which was widely used search engine technology later. Advanced search technologies work for us every day to filter relevant resources and to deal with the impacts of information overload. This article brings together findings on an research of *human-agent interaction*. Below we focus on information agents and some trends in general and following details on methods and key findings of our research are described.

6.2 Agents Working for Us

Research on agents has been ongoing since the beginning of 90's mainly in the specialized fields of DAI (Distributed Artificial Intelligence) and later Multi-Agent Systems. Specialized agents were developed and thereafter, the focus shifted towards the Agent Systems Modeling. The most recently agent technologies are perceived in the wider IT context. Some of the trends in the area are autonomous computing, web services and service-oriented computing, P2P computing, semantic web, ambient intelligence, and grid computing [3].

Agent is considered as intermediary system with specialized skills. From narrower point of view, agent is a software robot operating relatively autonomously. Agents are able to decide independently and to act in open and dynamic environment. Agents also react upon the changes and interact not only with other agents (building agent communities and society), but also with people. "Agents are a way to manage interactions between different kinds of computational entities and to get the right kind of behavior out of large-scale distributed systems" [2]. One of the key questions is how people interact with agents [3]. We tried to search for answer to this and other questions in our research project orientated on behavioral aspects.

More recently, agents are perceived as cooperating with people by solving problems with the use of "Internet-scale knowledge systems" [4]. These systems combine web 2.0 with intelligent agents and web services. Agents should serve as integrators of variable web 2.0 services which are relatively isolated and rapidly changing. Agents activities should result in creating specialized knowledge systems which provide access to integrated services using web 2.0 resources, methods, and tools. On the other hand, the idea of Internet-scale knowledge systems needs to be supported by the development of semantic web with the support of microformats. Microformats [6, 5] are simple open data formats created with the use of XML and XHTML (semantic mark-up languages). Microformats (like hCard, hCalendar, hReview, hResume, hAtom, XFN, rel-licence, XOXO etc.) enable decentralized development and the use of resources, services, and tools. Following, we would like to focus on specific type of intelligent information agents which were object of the research project.

6.2.1 Information Search Agents

Intelligent information agents provide access to information in distributed, heterogeneous, and dynamic environment. They usually integrate advanced functions of search, resource analysis, relevance evaluation, personalization, information products creation, etc. These systems collaborate with other agents to achieve common aims. Agents learn from users' behavior while they search for information, use, and evaluate it. According to user profile they do not only adapt to certain user needs but also support him/her to search for relevant information more effectively. Our research was aimed at testing and analysis of interaction of respondents with intelligent information search agent, Copernic Agent Professional (CA) [9]. Other well-known agent-based search tools are SearchBot or AllThat.

Copernic Agent Professional (Fig.6.1) is specialized for advanced web information search and provides many sophisticated functions. This search tool track for changes in web pages' content and queries, summarize information from resources, analyze search results, access invisible web etc. Despite of these advanced features, CA was too complicated for, especially novice, users. Research discovered interaction issues which enabled rethinking CA interface design.

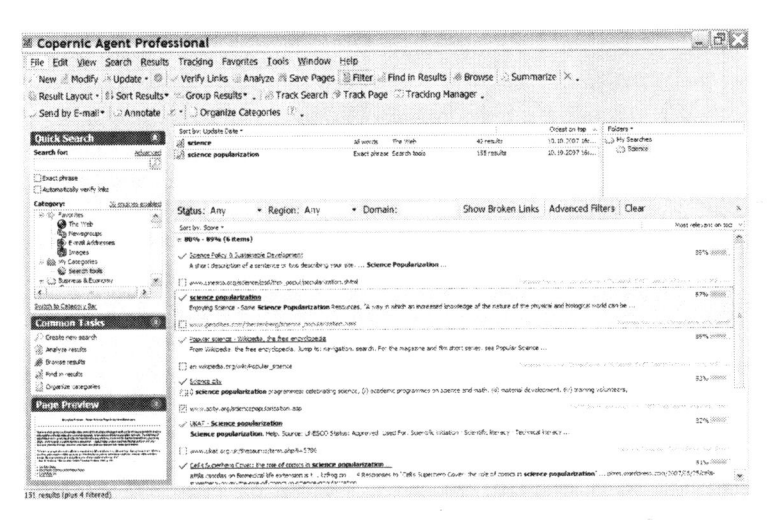

Fig. 6.1 Copernic Agent Professional interface

6.3 Research on Interaction of People with Intelligent Information Search Agent

Many problems associated with work efficiency of employees result from software tools with limited respect to usability. Usability is an important interactivity factor which takes effectiveness, efficiency and satisfaction with the product into account. Some of interaction barriers can result from reflecting only the thinking and models of developers into the design. We would like to describe our approach to analysis of interaction of people with Intelligent Information Search Agent.

6.3.1 Exploring Interaction

Our research [8] was conducted after completing pilot study in two phases starting with data collection and analysis (from October 2007 to January 2008) followed by synthesis (from January 2008 to February 2008). Research applied mainly qualitative methodological approach using semi-structured interviews, think-aloud method, and observation. Validation was enabled not only by triangulation of methods, but also by presence of three researchers, detailed description of methods, and informing respondents on research aims, data collection, and use. Analyses were also reviewed by respondents who gave us feedback and corrections (in interpretations) which were implemented. Collected data were recorded using digital recorder and screen capture application [9]. Additionally, data collection was also enabled by taking of observation notes. Combination of different data collection approaches were applied depending on the nature of the specific method (e.g. think-aloud data were captured by digital recording and screen capturing). All the data were transcribed from various resources and synthesized.

Respondents participating in the research were selected according to their internet information search experiences, and CA level of skills. Initial semi-structured interview was aimed at capturing data on identification of five respondents and their language skills, description of future plans, recent tasks and experiences (with technologies, internet information search and CA). Thinking aloud enabled collection of data on search, cognitive, and affective processes by the interaction. Respondents were expected to think aloud when solving tasks connected to information search and filtering, results' organization (grouping, sorting, folder categorization), summarization, source annotation, bookmarking and results' tracking. Concluding interview delivered details on problems by interaction with CA, interface limitations, level of satisfaction with search agent and proposals for its performance improvements. Observation notes were taken mainly by task completion (thinking aloud) and concluding interview when respondents went back to the interface to explain specific issues.

Consequently, collected data were transcribed and conceptually analyzed. Conceptual analysis was conducted manually according to the dimensions of identification, activity, approach, object, feature, time, and quantity. From analysis, tables with categorized concepts were derived. They were used as a basis for further analysis and hierarchical maps which were created with the use of mind mapping software [10]. Further analysis was aimed at cognitive and affective behavior by interactive information search. Synthesis resulted in interpretations, theoretical models [11] and practical recommendations for CA search tool performance improvements. Research outcomes are based on analytical results of cognitive and affective aspects of interactive information search behavior.

6.3.2 Research Outcomes

According to our research outcomes, cognitive and affective behavior play significant role in interaction of respondents with CA. Driving power of overall search effectiveness were *affective behavioral aspects*. Uncertainty, discontent and perplexity were the most frequently present negative emotions. On the other hand, research confirmed that satisfaction, interest and curiosity can also have strong influence on interaction by information search. Recommendations for CA search tool performance improvements were derived from analysis of affective behavior and answers of respondents from concluding interview aimed at discovering problems, CA limitations, and level of their satisfaction with interaction. Ideas for improvements were proposed by respondents and also included in the recommendations.

Further, we would like to introduce some of the crucial interaction problems together with recommendations (more detailed analysis could be found in [11]). These findings can be relevant not only for information search tools' developers, but also for wider software engineering and HCI community.

The most problematic aspect of the interaction with CA is application's complexity which leads to the confusion of respondents. Too many options and functions provided in the menu need to be simplified. Advanced search options are not easy to find and respondents also pointed out simple search was not very visible for them at first sight. On the other hand, respondents were disappointed by simplicity of advanced search form, because they expected more options (similar to common search engines as Google, etc.).

Respondents also had problems to understand the way search results are saved. CA provide search history feature which is difficult to differ from regular search list. Another usability problem associated with search results was that not visited links are not well distinguished from visited. Also number of search results was not present at the screen all the time.

Respondents had significant problems with understanding of features and their names. They were confused by following functions: Annotate/Summarize, Track search/Track page, Sort/Group, and Add folder/Save page. Functions should be explained with the support of help.

CA provides categories of search tools organized to folders according to the topic. Respondents perceived that they couldn't add new search tools to the categories as a very limiting. CA needs to open to its environment and provide users more flexibility.

Level of satisfaction with interaction was influenced by success in task solving, information need satisfaction, search results' utility in problem-solving, and CA features. Satisfaction was also affected by cognitive aspects, e.g. topical knowledge, knowing of the results, previous search experience, habits, known procedures, etc. All the respondents but one made assumption on their satisfaction with CA functions. Despite of this fact, they emphasized interaction problems and many CA limitations when thinking aloud by task-solving.

Respondents made proposals of improvements on the level of CA application and its specific functions. According to them, CA could be web-based application with advanced functions providing personalization and customization. They would layer the functions which could be added continuously not all at once. It was also proposed to open the source code of CA so it would be accessible to wider developers' community for creating e.g. add-ons. Respondents also recommended simplifying folder organization of results and search tools' categories. They found better to rename some function so they would be more understandable. Filtering, especially advanced filtering options, could be more visible. More respondents indicated they would appreciate back feature to undo specific actions easily. Further details on recommendations for CA performance improvements and other research outcomes were published and could be found in previous works [8, 11].

6.4 Conclusion

Ease and effectiveness of information technology use play important role in the context of workplace and organization. Advanced search tools are designed to decrease impacts of information explosion. According to our research outcomes, it doesn't need to apply with any exceptions. Introduced methodology overlaps the borders of CA search tool research and can be applied to test and analyze interfaces of other information technologies and tools. We believe that other fields could also benefit from described research methodology and outcomes.

References

1. Bush, V.: As we may think. Atlantic (1945),
   http://www.theatlantic.com/doc/194507/bush
2. Luck, M.: Best-kept secret agent revealed. Computer Weekly (October 12, 2006),
   http://www.computerweekly.com/Articles/2006/10/12/219087/best-kept-secret-agent-revealed.htm
3. Luck, M., McBurney, P., Sherory, O., Willmott, S.: Agent technology: Computing as Interaction: A roadmap for agent based computing. Agentlink III (2005),
   http://www.agentlink.org/roadmap/al3rm.pdf
4. Tenenbaum, J.M.: AI Meets Web 2.0: Building the Web of Tommorow, Today. CommerceNet Labs Technical Report 05-07 (December 2005),
   http://wiki.commerce.net/images/a/a2/CN-TR-05-07.pdf
5. Allsopp, J.: Microformats: empowering your markup for Web 2.0. Friends of Ed (2007)
6. Microformats, Last update (May 27, 2009), http://microformats.org
7. Copernic Agent Proffesional,
   http://www.copernic.com/en/products/agent/professional.html
8. Greskova, M.: Cognitive foundations of information science: interaction of human with information search agent [PhD thesis], Comenius University, Bratislava, FFUK (2008)
9. CamStudio, http://camstudio.org
10. FreeMind, http://freemind.sourceforge.net
11. Greskova, M., Steinerova, J.: Relevance in the light of interactive revolution: interplay between human and information search agent. In: Human-computer interaction and information services, pp. 1–14. ÚISK, Praha (2008)

Chapter 7
The Use of Natural Language as an Intuitive Semantic Integration System Interface

Stanisław Kozielski, Michał Świderski, and Małgorzata Bach

Abstract. This paper describes the need for intuitive interfaces to complex systems that take their origin from the concepts of Semantic Web. Then it shows how Semantic Integration System HILLS can benefit from being merged with Pseudo Natural Language layer of Metalog system. The cooperation of these two systems is not perfect though - second part of the paper shows guidelines to its improvement in context of Polish language.

7.1 Introduction

Typical search for an information in the Internet, for example by means of popular search engines, requires an ability to interpret and associate information, especially for more complicated queries. As an example of such complex queries one may see a search for the information coming from many sources. To enable automatic interpretation of data placed in the Internet, Tim Berners-Lee proposed the idea of Semantic Web [1]. The foundation of this idea is to add semantic description to the information accessible in the global Internet, what led to development of languages and technologies of formal knowledge representation: RDF, RDFS, OWL languages, ontology engineering, tools and decision support systems.

The ontological descriptions of information sources, which appear more and more widely in the Internet, form the basis for building intelligent applications, capable of searching, understanding and integrating information from many sources. The problem, which arises though, is communication of such complex applications with a user. Generally, the best form of such communication is natural language.

Stanisław Kozielski · Michał Świderski · Małgorzata Bach
Silesian Technical University, Gliwice, Poland
e-mail: Stanislaw.Kozielski@polsl.pl,
e-mail: Michal.Swiderski@polsl.pl,
e-mail: Malgorzata.Bach@polsl.pl

E. Tkacz and A. Kapczynski (Eds.): Internet – Technical Development and Appl., AISC 64, pp. 51–58.
springerlink.com © Springer-Verlag Berlin Heidelberg 2009

This paper takes up the aforementioned problem. First we present the Metalog [2] system and we introduce our semantic integration system that integrates data from many Internet sources. Then we show how our system was merged with the Metalog, what gave us, among others, possibility of communication with a user by means of the language similar to the natural language. Because Metalog is restricted to English, in the next step we show how to adapt our system to the Polish language.

7.2 Metalog

In our times computer systems are becoming more and more indispensable in many fields of human life. Using them is often necessary for the people who have little or no experience with computers. For these people mastering formalized language of communication with the complex systems is very difficult or even impossible. The best solution would be to equip the systems with the capability to communicate with the user in natural language. Although research on linguistic analysis has been conducted for decades, a complete (full) analyzers have not been developed for any natural language (even for English). Some scientists have even suggested that natural languages are too complicated and ambiguous to create a computer system being able to analyze and generate the whole variety of meanings [3]. Because of this we may use "pseudo" natural language. Pseudo - means that the natural language that is used is not the complete natural language that is used by all of us, but a subset of it. We can limit the language to a simpler grammar forms. The example of such system (Metalog) is presented below.

Metalog tries to meet two common necessities in the Semantic Web: the ability to reason in the Web, on the one hand, and the ability to bring these advanced technologies to the widest possible audience, on the other hand. The Metalog is composed of three fundamental components. The first component is the Metalog Model Level (MML) which extends plain RDF with a "logical layer", enabling to express arbitrary logical relationships within RDF. The second component is the Metalog Logic Level (MLL), that expresses the semantics of the MML in Prolog logic programming language, which is a superset of Datalog, used as a query language in the HILLS system (see 3.1). The third component is a PNL user interface, which is totally unambiguous, and it achieves this by limiting greatly ways the sentences can be written.

Metalog programs are composed of three kinds of sentences: the representations, the assertions and the queries. Representations associate some word to its corresponding representation, e.g.:PETER represents the person "Peter Smith" from the company "http://www.example.com/staff". Assertions are the sentences where we state something. Basically these sentences have simple structure: *"subject predicate object"*. For example: PETER IS "engineer" indicates that *PETER* is the subject of the sentence, *IS* is the predicate (verb), and *"engineer"* is the object of the sentence.

Assertions can also be much more sophisticated. They can contain logical operators or they can express deduction rules, e.g.: `if SHE has a "degree" in "computer science" then SHE "is" "smart"`. Queries are sentences that end with a question mark. For instance, to answer a query: `do you know whether PETER IS SOMETHING?` the logic programming engine will generate a set of values that bound to a variable *SOMETHING* will satisfy the logic program.

As we can see in above examples, names (literals, strings) are written in Metalog in a simple way, by enclosing them within double quotes. Words that are in lower case are keywords (reserved words), or if not, they are ignored. Variables are expressed using names all in upper case.

Metalog supports three main classes of reserved keywords corresponding to logic conjunction ("and"), disjunction ("or") and implication ("then", "imply", "implies"). A full description of all the Metalog keywords is not within the scope of this paper, we mention only that Metalog has also support for basic arithmetic and comparison operators.

In the next chapter we presented how the Metalog system was used with semantic integration system - HILLS.

7.3 HILLS Semantic Integration System

The aim of a semantic integration systems is to integrate the data exposed by different Internet data sources, which are by the rule heterogeneous. To fully integrate data, we have to see the process of integration on three levels of abstraction: syntactic, structural and semantic [4]. *Syntactic integration* deals with reconciling different data types, *structural integration* is about reformatting heterogeneous data structures to a homogeneous one, whereas *semantic integration* is about a proper identification of data semantics and performing integration on its basis. The HILLS system (*Hybrid Lightweight Local-As-View Integration System*) [5] removes semantic heterogeneity by formal semantic description of a user query and each data source with the use of ontology. By ontology we mean the specification of a conceptualization [6] or more verbosely: formal, explicit specification of conceptualization of a certain domain.

7.3.1 HILLS Architecture

The HILLS system architecture builds upon the standard LAV (*Local-As-View*) [7] semantic integration system architecture. We present HILLS architecture on Fig. 7.1 and then we briefly describe its main components as well as steps of user query evaluation.

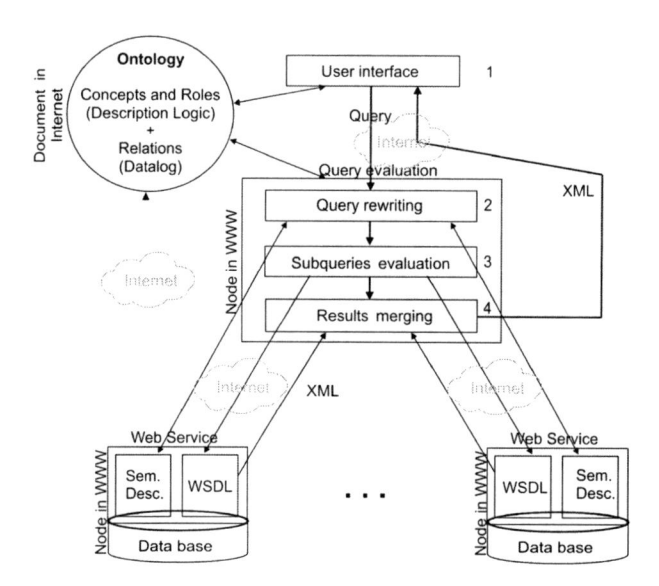

Fig. 7.1 The HILLS system architecture

Ontology

The heart of the HILLS system is the Ontology[1]. The user query and the data sources descriptions are built with use of the Ontology vocabulary. The Ontology has a hybrid structure and consists of two components: (1) relational, represented by *n*-ary relations, and (2) terminological, represented by Description Logic [8] axioms. For the terminological component's sake a special Description Logic language has been designed: DL-SI (*Description Logic for Semantic Integration*) [5]. The DL-SI is asymmetric i.e. it allows different sets of operators to appear on left and right hand side of a subsumption axiom. On both sides we allow: A, $\top$, $\bot$, C $\sqcup$ D, $\forall$R.C, and additionally on right-hand side of subsumption axiom we allow: $\neg$A, C $\sqcap$ D, $\leq$ n R, where A stands for atomic concept, C and D for complex concepts, and R for any role. The above mentioned constraints are proven to ensure the soundness and completeness of the query evaluation process.

In the example below relational component of the Ontology is empty and the terminological one is as follows:

```
area ⊑ ≤ 1 boundary
city ⊑ area ⊓ ≤ 1 name
recreationArea ⊔ industrialArea ⊔ publicArea ⊔ privateArea ⊑ area
recreationArea ⊑ ¬ industrialArea
publicArea ⊑ ¬ privateArea
bodyOfWater ⊔ park ⊑ publicArea
factory⊑ industrialArea
```

---

[1] Aka *Conceptual Schema, Mediated Schema* or *Global Schema*.

The concept area has at most one *boundary*, *city* has at most one *boundary* and one *name*, *recreationArea* and *industrialArea* are disjunctive, and so are *publicArea* and *privateArea*.

User Query

The user query is formulated in Datalog with the Ontology vocabulary, where concepts are represented by unary predicates, roles by binary ones and relations are used in a straightforward manner. If we exclude recursive queries, we can evaluate the Datalog query with use of relational algebra, SQL and relational databases [9], because not recursive Datalog corresponds to SQL without grouping and aggregation. Instead of Datalog formal definition we present the comparison of basic queries in Datalog and SQL. We use a simple database with only two relations: employee(e_id, name, surname, age) and child(e_id, c_id, age). We can observe in

Table 7.1 Comparison of SQL and Datalog

Query Type	SQL	Datalog
Projection	`SELECT name FROM employee`	`query(name):- employee` `(e_id, name, surname, age).`

Selection	`SELECT * FROM employee`
	`WHERE    age > 30`
		`query(e_id, name, surname,`
		`age)  :-  employee(e_id,`
		`name, surname, age), age`
		`>30`

Natural join	`SELECT e.e_id, name,`
	`surname, e.age, c_id, c.age`
	`FROM employee e, child c`
	`WHERE e.e_id = c.e_id`
		`query(e_id, name,`
		`surname, e_age, c_id,`
		`c_age)  :- employee(e_id,`
		`name, surname, e_age),`
		`child(e_id, c_id, c_age).`

Theta join	`SELECT e.e_id, name,`
	`surname, e.age, c.e_id,`
	`c_id, c.age FROM employee`
	`e, child c WHERE e.age <`
	`c.age`
		`query(e_e_id, name,`
		`surname, e_age,`
		`c_e_id, c_id, c_age):-`
		`employee(e_e_id, name,`
		`surname, e_age),`
		`child(c_e_id, c_id, c_age),`
		`e_age < c_age.`

Table 7.1 that in Datalog: (1) we can freely choose variable names as values bound to them depend on their position in predicates, (2) we acquire join by the presence of the same variable in two predicates. Additionally in HILLS system we allow the use of a member selection operator '.' (originally absent in Datalog), similar to the one known from object oriented languages, in order to access parts of XML documents bound to variables.

We can now ask a query for a recreation area that is at the same time a public area and is located in London (*within* is a built-in spatial operator in HILLS):

```
q(a, c) :- recreationArea(a), publicArea(a), city(c),
           within(a.boundary, c.boundary), c.name = 'London'
```

We can also specify five data sources that are described as views over the Ontology:

```
v1(x)  :- recreationArea(x).
v2(x)  :- city(x), x.name = 'London'.
v3(x)  :- bodyOfWater(x).
v4(x)  :- park(x).
v5(x)  :- factory(x).
```

The data sources are modeled as Web Services, and their semantics is specified by a predefined web method exposing semantic description in an XML format based on RuleML [10].

Query Rewriting, Evaluation and Results Merging

After the construction, a user query is checked against Ontology for satisfiability. If the query passes the test, meaning that a query has a chance to retrieve some data, it undergoes a complex process of a query rewriting, which aims to rewrite a query from an Ontology vocabulary to references to specific data sources. The HILLS system uses modified Destination-Based Algorithm (original [11], modified [5]) which takes into account: user query, Ontology and data sources. Continuing previous examples, modified DBA generates two rewritings:

```
q(a, c)  :- v1(a), v3(a), v2(c), within(a.boundary, c.boundary),
            c.name = 'London'.
q(a, c)  :- v1(a), v4(a), v2(c), within(a.boundary, c.boundary),
            c.name = 'London'.
```

After successful completion of a query rewriting step, the HILLS system proceeds to evaluate rewritten queries and merge received results. It is achieved by the following actions: (1) each rewritten query is divided into as many subqueries as many data sources it refers to, (2) the subqueries are evaluated against the data sources by remotely invoking their web methods, (3) received partial XML results are stored in XML-enabled data base, (4) rewritten queries are translated into SQL/XML form and are evaluated in an XML-enabled data base and (5) the merged results are passed back to the user interface as an XML document(s).

7.3.2 HILLS and Metalog

In the previous chapter we introduced the Metalog system, which does exactly what HILLS system lacks: provides an intuitive PNL interface and translates user input into logic programming clauses. To add Metalog PNL capabilities to HILLS system we had to solve a number of minor problems: (1) two systems need to communicate with each other, (2) concepts and *n*-ary relations are not present in Metalog, (3) a member selection operator '.' is not present in Metalog and (4) Ontology vocabulary must be presented in Metalog in an intuitive manner. These problems are addressed in following paragraphs.

The Metalog system cooperates with the Prolog engine as a logical backend. The communication between them is organized via files and scripts executed when files are ready on disk. We reuse this mechanism: we capture these files, translate into

HILLS format (if PNL is transformed into Datalog subset of Prolog) and treat them as ordinary HILLS user queries.

Roles from HILLS Ontology correspond directly to predicates in Metalog, but in Ontology there are also constructs which are not originally present in Metalog: concepts and n-ary relations ($n>2$). All predicates in Metalog are binary, while concepts in Description Logic correspond to unary predicates. This is not a serious problem as we can introduce syntactic shortcut in Metalog to present concepts: for each concept we introduce a separate constant of its name (e.g. *CONCEPT*); we also introduce a predefined binary predicate *IS(individual, CONCEPT)*, that is translated into communication file as *CONCEPT(individual)*. We could also handle n-ary relation in a similar manner, but unfortunately Metalog's PNL grammar is not prepared for such an extension and that's why we restrict the current prototype to unary and binary predicates.

Another problem is an absence of a member selection operator '.' in Metalog and its presence in HILLS. To solve this issue we add a keyword "*of*" to Metalog which works as follows: (1) to indicate a property of an object we write in PNL "*PROPERTY of OBJECT*"; (2) when such text is found during parsing we remove the "*PROPERTY of*" from the PNL and when the query is being written to the file we replace the *OBJECT* with a property selection of this object: *OBJECT.PROPERTY*.

The last problem is about presenting Ontology vocabulary in the Metalog in an intuitive manner. We achieve this by the use of Metalog Representations which allow to describe Ontology concepts and roles verbally and to create their upper case representations thus removing the need to use their names in double quotes.

Continuing our example, we can now present original user query in HILLS PNL:

```
do you know an AREA that IS a RECREATIONAREA,
and at the same time this AREA IS a PUBLICAREA,
and a BOUNDARY of this AREA lays WITHIN a BOUNDARY of MYCITY,
and MYCITY IS a CITY, and the NAME of MYCITY equals "London"?
```

7.4 Polish Language and Metalog

Metalog originally utilizes English, but using our previous experience [12][13] we tried to adapt Metalog to Polish language. The first (naive) attempt was to replace English words with their Polish equivalents. This approach didn't give satisfactory results as Polish unlike English is a language which inflects.

The second step was to eliminate the need to define many inflected forms for all variables. We had to apply the morphological analyzer which for every word gives its basic form (e.g. for the noun it is nominative in the singular, for the verb - infinitive, for the adjective - nominative of the masculine gender). For example, the sentence: MIASTO JEST "duże" (CITY IS "big") is changed by the morphological analyzer, to: MIASTO BYĆ "duży" (CITY be "big").

Another problem in the Polish is the free order of the words in sentences. It causes that the classic grammatical model of basic sentence, where we have a subject followed by a predicate and then an object (triple "subject predicate object"), isn't kept. Therefore, at present, we try to use semantic analyzers. Semantic analysis was realized by using Fillmore's case grammar. In this approach semantic cases express the semantic role played by each word (or words' group) in sentence (e.g.: in the sentence: `"Peter reads a newspaper"` noun *"Peter"* is a semantic case *AGENT*, verb *"read"* plays the semantic role *ACTION* and noun *"newspaper"* is *OBJECT*). Such semantic structure will facilitate the mapping of words to appropriate elements of the triple. The work on using the semantic analysis in Metalog is in progress.

7.5 Summary

In this paper we described how HILLS system can benefit from being merged with Pseudo Natural Language layer of Metalog system. The cooperation of these two systems requires a few adjustments, but allows to formulate queries to HILLS system in an intuitive manner. In the second part of the paper we showed preliminary results of accommodating Metalog to Polish language. We plan to follow that direction in the future, so that our prototype may incorporate a fully-fledged Polish PNL interface.

References

1. Berners-Lee, T., Hendler, J., Lassila, O.: The Semantic Web. Scientific American 284(5) (2001)
2. The Metalog Project, http://www.w3.org/RDF/Metalog/. W3C
3. Dreyfus, H., Dreyfus, S.: Mind Over Machine: The Power of Human Intuition and Expertise in the Era of the Computer. The Free Press (1998)
4. Visser, U., Stuckenschmidt, H., Schlieder, C.: Interoperability in GIS - Enabling Technologies. In: 5th Conference on GIS, Palma (2002)
5. Świderski, M.: Semantic integration of geospatial data sources in the Internet. PhD dissertation, Gliwice (2007)
6. Gruber, T.: A Translation Approach to Portable Ontology Specifications. Knowledge Acquisition 5 (1993)
7. Levy, A.Y.: Logic-Based Techniques in Data Integration. Univ. of Washington (1999)
8. Baader, F., McGuinness, D., Nardi, D., Patel-Schneider, P.: The Description Logic Handbook: theory, implementation, and applications. Cambridge University Press, Cambridge (2003)
9. Halevy, A.: Answering Queries Using Views: A Survey. VLDB Journal 10 (2001)
10. Boley, H., Grosof, B., Tabet, S.: RuleML Tutorial,
    http://www.ruleml.org/papers/tutorial-ruleml-20050513.html
11. Wang, J., Maher, M., Topor, R.: Rewriting general conjunctive queries using views. In: Australasian conference on Database Technologies, Melbourne (2002)
12. Bach, M.: Methods of Constructing Tasks of Searching in Databases in the Process for the Translation of Queries Formulated in the Natural Language. PhD dissertation, Gliwice (2004)
13. Bach, M., Kozielski, S., Świderski, M.: Application of Ontology to Relational Database Semantics Description in Context of Natural Language Queries Analysis. In: BDAS Conference, Ustroń (2009)

Chapter 8
Student's Electronic Card: A Secure Internet Database System for University Management Support

Andrzej Materka, Michał Strzelecki, and Piotr Dębiec

Abstract. In the paper, design concept, architecture and functionality of the "Students' Electronic Card" computer system are presented. The system was developed in 1998-2007 and deployed at the Faculty of Electrical, Electronic, Computer and Control Engineering, Technical University of Lodz, Poland. It aids to efficiently organize activities of the community, including 4500 students along with 450 faculty and staff members. The core elements of the system are internet database (Oracle, Postgress) and a network of different functionality terminals (class assignment, cost analysis, marking, student portal, etc.). These terminals are used to enter, edit or retrieve data related to the management of teaching, research and other activities at the faculty, including funds allocation and cost analysis. A number of users of different roles and privileges are defined in the system. Smart cards are issued to users to store their personal secret keys for data encryption and electronic signature generation while communicating with the database. The system has been continuously updated and expanded for more than 10 years. One of the recent projects in this area is "Electronic Education Connectivity Solution", aimed at providing support to Bologna processed-governed student mobility between European universities, funded from 7FP UE program.

8.1 Introduction

While the clear mission of a university is education, it is certainly a very large business enterprise, entailing huge administrative activity [1]. From a managerial point of view, the university campus is a rather large and complex business organization. It has to provide many diverse services to a large number of people who are engaged

Andrzej Materka · Michał Strzelecki · Piotr Dębiec
Technical University of Łódź, Institute of Electronics, Wólczańska 211/215,
90-924 Łódź, Poland
e-mail: {materka,mstrzel,pdebiec}@p.lodz.pl

E. Tkacz and A. Kapczynski (Eds.): Internet – Technical Development and Appl., AISC 64, pp. 59–72.
springerlink.com

in a huge variety of activities [2]. The services have to operate smooth to the satisfaction of their users; they should help, not disrupt, the statutory user activities and they should be cost-effective [1].

The core of campus services is now a database system or a network of local databases, seamlessly connected to each other. Most of the information stored includes user individual data, such as personal information or marks acquired by students. Such sensitive individual information should not be disclosed to unauthorized persons. Thus users transfer the information to the database and read it from the database through secure communication channels. To make the channel secure, the user must be identified by a unique ID code, used e.g. to encrypt the information and/or attach electronic signature to the data chunks [3], [4].

Many campus services include payments, e.g. for parking, printing, or operating vending machines. Cashless purchase is definitely preferred to cash payments, as it reduces the costs of handling and moving cash. Smart card used as an electronic purse [5], [6], [7] provides a means of eliminating cash flow from campus transactions. Yet another vital issue relates to access control - to rooms, facilities, libraries, computer networks. To be authorized for a service, the user has to prove his/her identity or present a certified permit which can be verified in the system by reading information stored in a campus card memory.

New opportunities and challenges are generated by the process of integration of European educational systems [8]. The idea of student and staff mobility between universities has gained a wide acceptance as expressed by the Bologna Declaration [9]. To implement this idea in practice, one needs a common ID number that could be stored in a campus card memory, along with the study-relevant personal information, to be carried by a student to any university he/she is actually enrolled to.

All the examples mentioned above indicate the need for a secure information storage that can be carried by its user/owner everywhere. Over the years, a plastic card has become such a storage device. Depending on application, a magnetic stripe or an electronic memory/microprocessor chip is embedded in the card. The cards equipped with a microprocessor, and an operating system, are called "smart cards". Unlike magnetic stripe cards, smart cards carry all necessary information and processing functions on the card. Therefore, they do not require access to remote databases at the time of the transaction [10]. The cost of a single, non-personalized card may range from 0.2 eurocents for a magnetic stripe, through 2.50 euro for a memory card to 10 euro for a smart card. There were over 900 million credit cards (of magnetic stripe type mainly) in circulation in 2006, global production of smart cards surpassed 5 billion in 2008 [11]. Major uses include providing enhanced financial services, increasing the security and flexibility of cellular phones.

It is evident that plastic cards have far-reaching impact on e.g. financial services, transportation, telecommunications, and health. A smart card is now in fact a wearable computer that serves as a secure storage of personal information in almost every place and time of our activity.

The smart card is of course not a panacea for all user authentication problems. Since the card can be stolen together with a PIN code to it, then in high security applications there is a need for other, supplementary techniques of people identification.

Most of these techniques are based on measuring personal characteristics of user body or behavior, such as fingerprint, retina blood vessel patterns or voice parameters. These types of biometric information can be used to enhance (augment) or replace information stored on the card. Development of the biometric authentication/identification instruments integrated with campus card systems requires new sensors for reliable measurement of personal information and new methods of signal processing, image analysis and pattern recognition. These issues again fall within the range of interest of the Students' Electronic Card system designers and developers [12].

This paper provides information about architecture, functions and implementation issues of the Students' Electronic Card system. The system is an extensively verified, working example a vital, widespread and still growing application of secure internet database systems designed for the support of management of a very complex enterprise - a university.

8.2 Project Background

The idea of applying electronic cards to store student marks and help in the management of teaching at the Faculty of Electrical, Electronic, Computer and Control Engineering, Technical University of Lodz (EECCE TUL) was first considered in 1996-98. The need of introducing electronic documentation was well recognized at the Faculty at that time. The four-fold increase of the number of students in a few years (to about 4500 students, which made the EECCE the largest TUL faculty), caused many organizational problems to academic and administrative staff. These problems endangered quality of student services and needed quick solution. Initially, memory cards were considered [1] to serve as a form of electronic student book of records. Each student was expected to be issued a plastic card with her/his photo for visual identification, and a digital memory of capacity sufficient to store all data during the whole study. This card was to be carried in a pocket of the paper book of records (which can not be eliminated due to national regulations). Teachers were to write in the marks both to the paper book and to the card memory, the latter with the use of a card reader connected to a computer.

Prototype card readers and software modules were designed, developed and used for tests which confirmed feasibility of the system. The design was based on Siemens cards [13]. After assessment of the investment, material and exploitation costs [1] of the enterprise, accounting for the size of the Faculty (4500 students and more than 450 faculty and administrative/technical staff) the basic assumptions concerning the system specifications were modified. Giving each student a card for marks duplication and further off-line storage in the system was considered too costly. A decision was made the marks will be stored in a common internet database accessed by teachers and administrative staff. The microprocessor cards would be issued to teachers and administrative staff only, in the first phase of system deployment at the Faculty. Main function of the card was the storage of user identification data and the private keys used for encryption and digital signature of the data exchanged between the internet database and the system terminals. Based on this decision, a comprehensive faculty information processing system has been developed and expanded, now involving the students also as passive users (i.e. with no authorization to write the crucial data into the base, yet). It is expected that wide introduction of electronic student ID card [14] will allow granting the students the right to write the data, e.g. to electronically various forms, make payments, etc.

In general, language and legal issues have severe impact on university management system design. For this reason, there are no universal, comprehensive, global solutions available. The systems are developed locally, and then adapted to the needs of information exchange with collaborating foreign universities. At the time the development of the SEC system started, the only solution available on the Polish market was HMS product by Kalasoft company [15]. The HMS allowed for collecting of student marks and aided the timetable planning. However, it did not support the ECTS framework. HMS was not an internet solution; it operated on local area network only. Currently, there is a number of software solutions which support different fields of activity of higher education institutions. Big companies such as SAP, Oracle and Microsoft offer their well-known ERP systems, more or less customized to the higher education requirements. Smaller software developers propose dedicated, integrated solutions, e.g. Sygnity Edukacja.CL [16], Kalasoft HMS Solution [15], Comarch CDN Egeria Edukacja [17], Partners in Progress Uczelnia.XP [18]. Some systems e.g. SUSZI [19] or USOS [20], have been developed by universities, or by university consortia. Comparative in-depth analysis concerning functionality and security of these systems has not been performed and published yet, to the authors knowledge. From the point of view of prospective system users, such an analysis is much needed. This paper reveals details of our system that can be useful for further comparative study.

The SEC system represents highly modular structure, it is easily upgradeable, and can be connected to computer systems already existing at the university. The flexible architecture of the SEC allows adaptation both to the needs of the particular university and to changes in law and regulations. The SEC features wide functionality, including class assignment, cost analysis, marking, student portal, teaching and research management. The developers of the SEC – experienced academic teachers and software engineers – are at the same time the system users. This results in a well-thought-out system functions and structure, as well as the user-friendly interface. The implemented algorithms of data encryption and electronic signature meet the FIPS standards issued by NIST.

The aim of this paper is to describe functionality and architecture of the system developed at the Faculty for more than 10 years now. Its publication will help sharing experience by universities implementing such systems at their campuses. Since design, development and implementation of campus card systems are very complex tasks, we believe such information exchange is a necessary component of this enterprise. There is strong Europe-wide commitment to this approach, with emphasis on developing standards for information exchange, especially in view of the huge diversity in tradition of European academic institutions [21].

8.3 Functionality and Architecture of the System

The architecture of the system, with a superimposed diagram of teaching information flow is illustrated in Fig. 8.1. All the information collected in the Faculty is stored in a single Internet database. The Oracle 10g database management system [22] was selected in the TUL EECCE computer "Students' Electronic Card" (SEC) system design. The SEC system users save the information to and retrieve it from the database by means of terminals – computers equipped with a card reader and specialized program. In the case of the management of information related to teaching, as illustrated in Fig. 8.1, there are 3 types of terminals – dean office terminal (TD), teaching unit terminal (TJD) and teacher terminal (TN). Other terminals types are used to collect and/or process information related to research, to the cost analysis of the Faculty activities, cards management, etc., as will be shortly discussed later.

A student portal, which is in fact a web application, is also integrated with the system shown in Fig. 8.1. Two secure interfaces/channels that synchronize two separate data sets have been developed. A set of data which undergoes frequent modifications (e.g. lists of obligatory and elective courses, partial and final marks, dean's decisions, financial data) is fetched online directly from the central Oracle database. The data undergoing occasional modifications (e.g. student and teacher personal data, curricula, contact details, etc.) is buffered in a Postgres database and synchronized twice a day to the main Oracle system. The two data sets have been carefully defined to always ensure the integrity of the information presented in the student portal. Students can browse through their academic books of records (marks, enrollment to consecutive semesters, fees due and paid, etc.) after logging in to a password protected portal. An important function of the portal is in collecting students' answers to questionnaires on subjects, teachers, and organization of studies. Thanks to the portal integrated with the information management system, more than 80% of the students give feedback on the important aspects of the study to the Faculty.

In the future, once electronic student ID cards are issued to the students, the functionality of the student portal will be extended to e.g. electronic submission of study-related documents to the dean office. At present, cards are issued to academic and administration staff. The cards serve for visual identification of their users and for the storage of information about the range of user authorization privileges in the system.

The access to the card is protected by a secret personal identification number (PIN). Data written in to the database are encrypted [3] and signed digitally [4], [23]. The private keys needed for these operations are stored in the card memory.

A system terminal is a standard PC computer run under MS Windows® operating system, equipped with a card reader and dedicated software. An example of a card reader connected to a notebook computer through a serial USB port is shown in Fig. 8.2. The application window seen at the computer display is a dean office terminal (TD) window instance. Software for the terminals was written in Object Pascal and C++ languages using Delphi and C++ Builder development

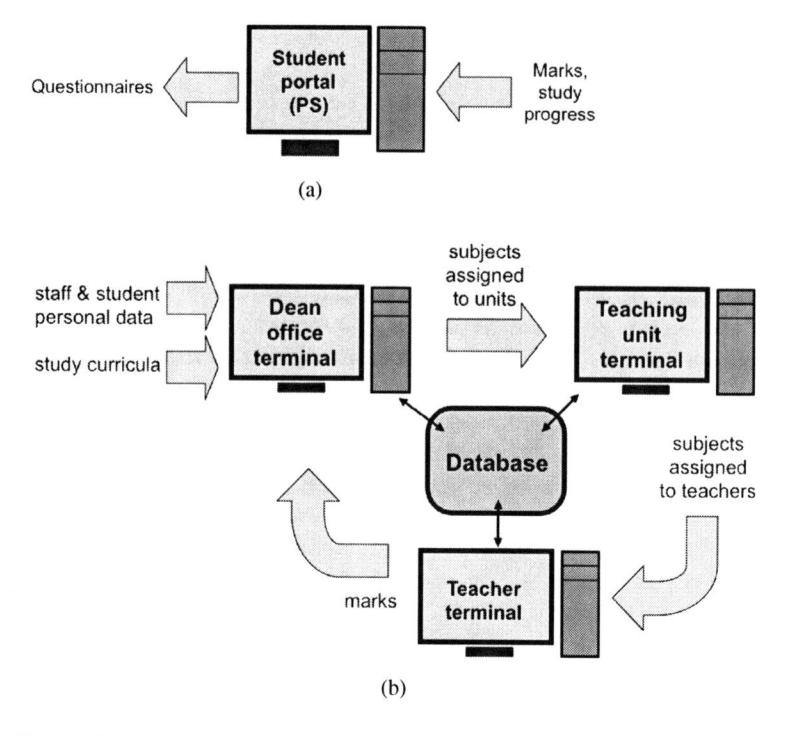

Fig. 8.1 SEC system architecture with an example of teaching-related flow of information

environment compilers [24], respectively. Operation of the student portal (PS) is encoded by PHP language [25] scripts run on an Apache server [26].

The electronic signature procedures are implemented by means of C++ and Object Pascal routines developed at the Faculty. The routines encode SHA-1 [27] and DSA [28] algorithms with the use of modular arithmetics [29]. Data encryption software modules use the TDEA (triple DES) algorithm [30].

Hybrid microprocessor cards Gemplus MPCOS-EMV R5 8000 (8kB) with Mifare contactless module (1kB) are currently used in the system [31], [32]. A special printer is applied to print the faculty and user-related information (Fig. 8.3) on both sides of the white plastic cards as they come from the vendor. A card issuing terminal (TWK) is used for card personalization. It comprises electronic encoding of

- data structure in the card memory,
- system information,
- user personal data,
- digital certificate of the card,

prior to graphic personalization.

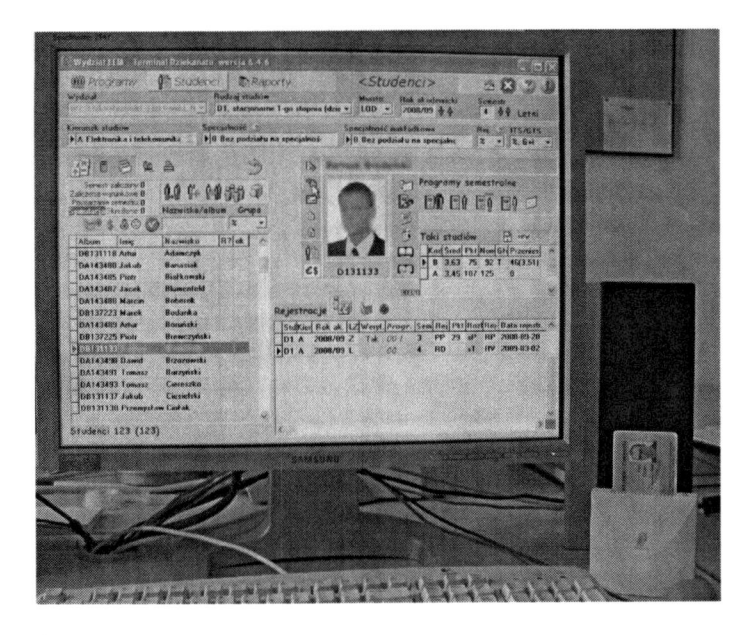

Fig. 8.2 Sample card reader (bottom right) connected to a PC terminal

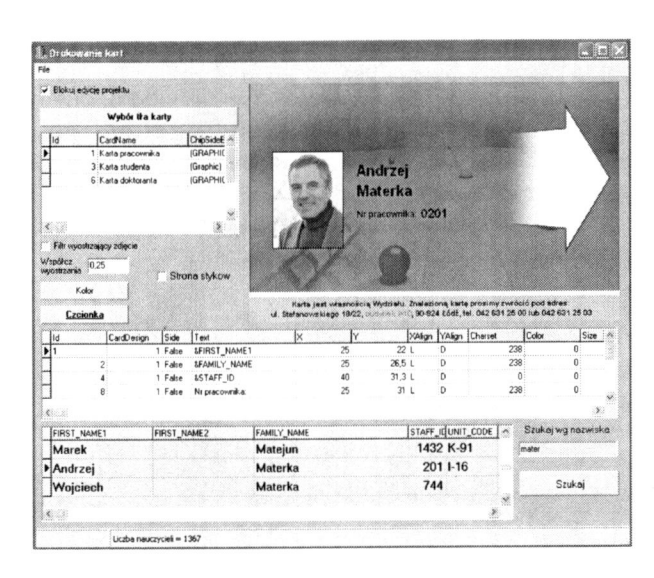

Fig. 8.3 Example window of card issuing terminal (in Polish: terminal wydawania kart – TWK)

8.3.1 Cards and System Security

The data protection means built into the SEC information management system include [31]

- limited access to data (user authorization required, many user types defined with limited access rights),
- protection of the data content from unauthorized reading (encryption),
- measures of checking data integrity and authenticity (digital signature).

The basic tool of information protection is the microprocessor card. Its internal resources are utilized to store a) user identification data, b) cryptographic keys, c) electronic purse files, d) access control information.

The MPCOS-EMV cards allow implementation of various protection mechanisms of the access to stored data. The right to perform any operation on the contents of the files – reading, writing, modification – can be limited. Similar restrictions can be applied to creation of new elementary files (EF) and dedicated files, i.e. folders (DF). The card operating system includes two tests that have to be passed to get rights to perform required operations. One relies on verification of passwords, stored in SCEF – secret code elementary files. The other is based on verification of a cryptographic key by using a challenge-response method [3]. In the latter case, the card microprocessor generates a pseudorandom number and sends it to the computer application which has requested creation of a new folder. The application encrypts the received number using the 3DES algorithms based on asymmetric key shared with the card and sends the encrypted message to the card. The card compares the received message with the one generated locally by its microprocessor. If the messages are the same, the card performs the requested operation; otherwise, an error message is sent to the application.

The four layers of access protection to the card data files, implemented in the TUL EECCE information management system are illustrated in Fig. 8.4. There are six user categories, each attributed different rights to access the card data files and database records – teachers, dean office administrative staff, dean and dean deputies, teaching unit terminal operators, system administrators, and students. Independently of the user category, their cards carry the following information:

- user and card ID data,
- login names to system applications,
- unique private DSA key (for digital signature),
- unique private DES key (for encryption),
- user categories assigned to the card owner,
- electronic purse structure/data,
- login names and passwords for access control of campus facilities,
- card certificate.

On top of the above-described numerous security means, protocols and procedures built into the system, standard firewall technique of terminal communication with the database is implemented in it. In general,

- the level of information protection depends on the estimated cost of information loss or modification,

Fig. 8.4 Layers of access protection to data stored in the card memory

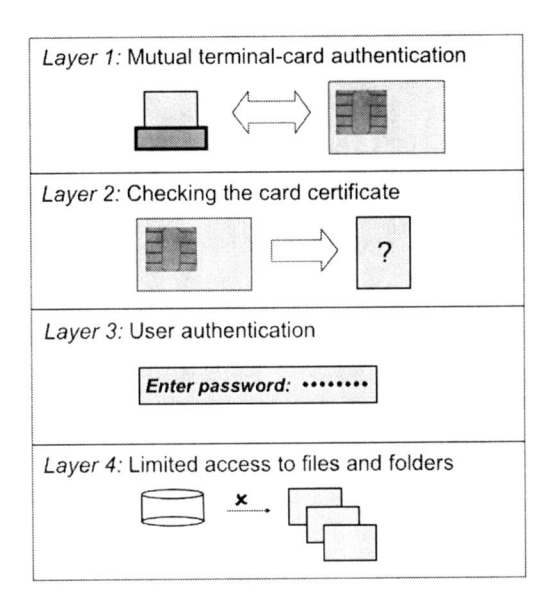

the security mechanisms available with the Oracle DBMS are utilized as well (e.g. database users with different roles and privileges),
- terminal communication with cards is kept secret,
- digital signatures allow continuous verification of data authenticity and integrity.

The use of electronic cards in the Faculty information management system led to a data security level at which data eavesdropping or modification by unauthorized users was not possible, in practice. For more than 10 years of the system operation in the Faculty environment, no unauthorised data access was observed. The single-internet-database architecture with terminals located in the Faculty teaching units (institutes and departments) allows for on-line assignment of teaching load and continuous verification of the assignment consistency throughout the whole Faculty. For example, information about the actual number of students who have taken any subject class are easily available to a dean once teachers make proper choices from a list of student names generated by the system.

After the pilot studies and initial tests, the system was put into operation in 2002, to include year 1 students, studying under European credit transfer system (ECTS) [33]. Year by year, newly enrolled student records were added to the database. Since 2006, data of all the Faculty students have been stored and processed by the system. The benefits of using the system, in terms of speed of information collection and data consistency, as well as in availability of information which would otherwise not be possible to extract, are very clearly seen now. This positive evaluation made the Faculty decide to expand the system functionality to cover other strategic areas of its activity, beyond course assignment and subject marking.

8.3.2 Data Synchronization with the University Database

Since the very beginning of the system development, and especially after introduction of the ECTS study system at TUL in 2002, it was evident that there is (and always will be) the need for data exchange between the University central database and faculty local systems. The principal rule is that there should be only one source of information present in the whole University system. For example, the student ID number should be issued in the central system and copied to the local systems. On the other hand, marks given to students for their study performance are collected at faculties and sent to the central database. In the ECTS system, these marks are aggregates (e.g. weighted averages) of the marks student achieve for exams, laboratory classes or tutorials of a given subject, carried out by Faculty units. These partial marks are not relevant to the operation of the central system where aggregate marks are important only, e.g. taken into account to pick up the best students who deserve a scholarship award. On the contrary, teaching costs are generated in the units who run partial classes (lectures, labs, etc.), so it is very important to record all of them in the faculty local database, e.g. for operational cost analysis.

Well defined protocols and procedures for data exchange between local faculty systems and the central database are essential to smooth, efficient operation of the whole university information management system. The data should be exchanged in electronic form, manual copying of information from one system to the other is obviously obsolete and ridiculous.

8.3.3 System Expansion

The database in a faculty system is a reach source of information, essential to various aspects of the faculty activity. Often, there is a need to extract this information in a non-standard form, e.g. for statistical purposes. A reporting module was designed and written for this purpose (Fig. 8.5). Various information-selection criteria can be selected from a list corresponding to the information fields stored in the database. The module dynamically generates an SQL query which is sent to the base. The result (a report generated in response) is displayed on the computer screen and can be saved as an HTML or Excel file. This unique flexibility in report generation is a very useful feature, e.g. to satisfy the needs of statistical reports preparation requested by the Ministry of Education.

One of the essential factors of faculty management is identification of various sources of costs of running the teaching process. The cost components (staff wages, cost of overtime hours) should be evaluated and compared with the faculty income (yearly funds granted by ministry, student fees, etc.). One can then compute average costs of running specific courses or form of studies, e.g. evening courses, part time studies, for cost efficiency. Such data are the basis for strategic analyses, e.g. in the area of employment structure that matches the teaching quality criteria and gives maximum cost efficiency at the same time. To allow cost analysis in the SEC

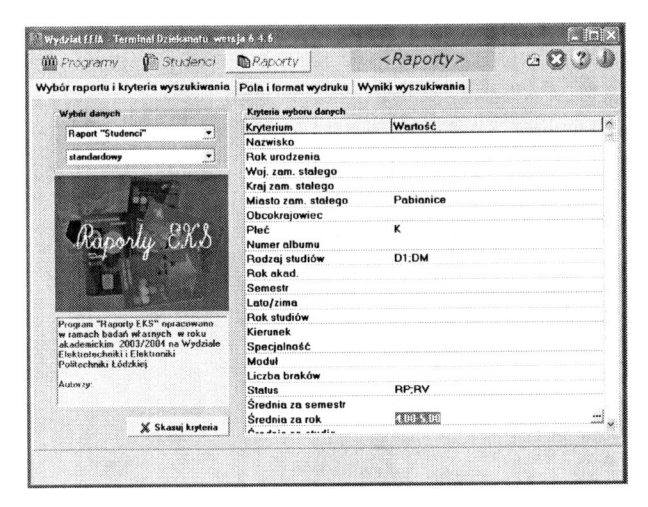

Fig. 8.5 Example window of flexible reporting program

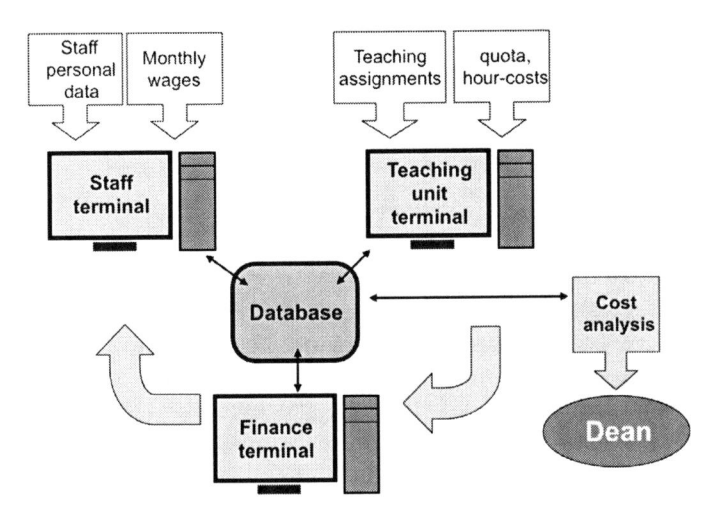

Fig. 8.6 Diagram of information flow for cost analysis

system, the system has been augmented with two terminals - staff terminal (TK with *Kadry* program) and finance terminal (TF with *Krezus* program), and the functionality of the teaching unit terminal (TJD, using the *Ekstazjusz* program) has been significantly enhanced. The terminals are used to collect and process information for the cost analysis task, as illustrated in Fig. 8.6.

The architecture of Fig. 8.1 was also adopted to apply the system for collecting and processing information related to the Faculty research activity. A new – research (TBN, Terminal Badań Naukowych in Polish) – terminal has been designed and implemented for the purpose, running the *Skryba* program. Depending on the user privileges (read-out from the user card), the terminal changes its interface to allow writing in the information to the database (for officers representing the faculty units, see Fig. 8.7) or reading out the information only (for dean office workers).

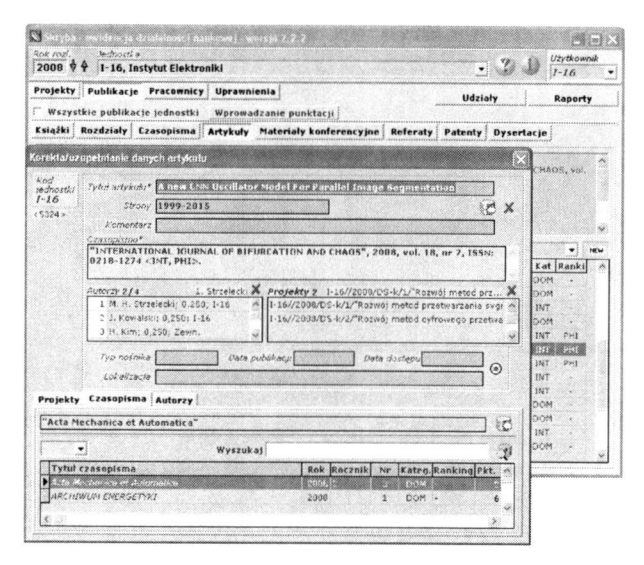

Fig. 8.7 Sample window for entering information on journal papers of TBN terminal

The TBN terminal allows collection of detailed information about research projects carried out in the institutes and departments, as well as on publications and various forms of research activity of all the staff members. Each form of research achievements is allocated a number of points which are counted to evaluate research activity in the faculty units. The points scored by a Faculty unit (institute or department) in a year are the basis for research funds allocation to it in the following year. Reports are generated, to make e.g. a list of publications for each unit, and the whole Faculty – for documentation required by the Ministry and Main Library, for staff assessment, planning, and other needs. A staff portal (based on an idea similar to the above described student portal) is under development now to provide individual staff members with an internet access to their research record data.

Another direction of the system expansion is in the area of card-based access control, e.g. to student laboratories. Both contact and contactless cards are considered in an in-progress Faculty project. Once electronic student ID cards are introduced within the University, they will be integrated with the existing system.

8.3.4 Design, Development, Update and Maintenance

The prototype system was designed and built in 1996-1998 by a 7-person team of experienced electronic engineers/computer programmers. Four of them had had previous 2-4 years teaching work experience at foreign universities (USA, Australia, Japan, Great Britain) which was important in terms of introducing the credit point study system at the Faculty. Its design and deployment required solution of many complex engineering problems (functionality specification, database design, software development - user interfaces, card reader communication, algorithms for data encryption and digital signature) and, even more difficult, many logistic problems related to organization of the work of different groups of users (teachers, administration, students). For 10 years of the system operation, its functionality has been continuously evaluated (by a formal team of nominated users), updated and upgraded by a team of a few programmers. One of the recent projects in this area is "Electronic Education Connectivity Solution", aimed at providing support to Bologna processed-governed student mobility between European universities, funded in 2009-2011 from Seventh Framework European program resources [34].

8.4 Summary and Conclusion

The core element of the faculty information management SEC system at TUL EECCE Faculty is a microprocessor card for the storage of user's personal information and private keys for data encryption and digital signature. The card provides high security level necessary for safe communication with a single database. Single source of information allows fast access to it and eliminates the very high costs of verification and processing of the data collected in a paper form. The introduction of the system contributed much to the quality of the Faculty work. Some vital decision-support analyses were not possible at all before, e.g. detailed cost evaluation of average teaching hour, say, at evening courses in satellite TUL campuses. Implementation of a campus card system is a complex task, not only from the technical, but mainly organizational point of view. We found it essential that users of the university systems were involved in the process of setting system specification, testing and evaluation of new hardware and software modules. It seams impossible that an external company, with no work experience in teaching and research organization can really succeed in development of a system that seamlessly integrates with faculty complex activities and helps in performing the faculty tasks rather than being a problem by itself.

Acknowledgments

The authors wish to thank Anna Janisz and Krzysztof Ślot for kindly supporting graphics design of some illustrations. Help provided by Marek Kociński in the paper type-setting is also much appreciated.

References

1. Allen, C., Barr, W.: Smart cards: seizing strategic business opportunities. McGraw-Hill, New York (1997)
2. Woodward, K.: Schools give smart cards a second look. Card-Technology 10, 45–48 (2005)
3. Hendry, M.: Smart card security and applications. Artech House (1997)
4. White, G., Fisch, E., Pooch, U.: Computer system and network security. CRC Press, Boca Raton (2000)
5. Zoreda, J., Oton, J.: Smart cards. Artech House (1994)
6. Guthery, S., Jurgensen, T.: Smart card developer's kit. Macmillan, Basingstoke (1998)
7. Frankl, W., Effing, W.: Smart card handbook. Wiley, Chichester (2000)
8. http://ec.europa.eu/education/higher-education/doc1290_en.htm (2009)
9. http://www.bologna-berlin2003.de/pdf/bologna_declaration.pdf (2009)
10. http://java.sun.com/javacard/reference/docs/smartcards.html (2009)
11. http://www.reportbuyer.com/banking_finance/debit_credit_cards/smart_card_market_forecast_2012.html (2009)
12. Ślot K.: Wybrane zagadnienia biometrii. WKiŁ (2008) (in Polish)
13. Goździk, M.: Microprocessor card systems and their application. Master's thesis, Technical University of Lodz, Faculty of Electrical and Electronic Engineering (1997) (in Polish)
14. Polish Ministry for Science and Higher Education: By-Law on documentation of the course of university study. Warsaw (November 2, 2006)
15. http://www.kalasoft.com.pl/index.php?id=5 (2009)
16. http://www.sygnity.pl (2009)
17. http://www.comarch.pl (2009)
18. http://www.uczelniaxp.pl/strona.aspx?id=55,7 (2009)
19. http://www.wszib.edu.pl/intranet/ (2009)
20. http://pl.wikipedia.org/w/index.php?title=USOS (2009)
21. http://www.ecca.ie (2009)
22. http://www.oracle.com/database (2009)
23. Schneier, W.: Applied cryptography, 2nd edn. Wiley, Chichester (1996)
24. http://www.embarcadero.com/products (2009)
25. Lecky-Thompson, E., Eide-Goodman, H., Nowicki, S.D., Cove, A.: Professional PHP. Wiley, Chichester (2004)
26. Laurie, B., Laurie, B.: Apache: the definitive guide, 3rd edn. O'Reilly, Sebastopol (2002)
27. FIPS PUB 180-3: Secure hash standard (2008), http://csrc.nist.gov/publications/fips/fips180-3/fips180-3_final.pdf
28. FIPS PUB 186-3: Digital signature standard (2009), http://csrc.nist.gov/publications/fips/fips186-3/fips_186-3.pdf
29. Cormen, T.H., Leiserson, C.E., Rivest, R.L.: Introduction to algorithms. MIT Press, Cambridge (1990)
30. FIPS PUB 46-3: Data encryption standard (1999), http://csrc.nist.gov/publications/fips/fips46-3/fips46-3.pdf
31. Ślot, K.: Karta elektroniczna systemu EKN. In: Proc. 1st National Seminar, Electronic University Staff and Student Cards K@Elektron, Cracow (2003) (in Polish)
32. http://www.gemalto.com/readers (2009)
33. EU: ECTS users guide. Brussels (2009)
34. http://www.eecscard.eu (2009)

Chapter 9
Video Shot Selection and Content-Based Scene Detection for Automatic Classification of TV Sports News

Kazimierz Choroś

Abstract. Visual retrieval systems need to store a large amount of digital video data. The new possibilities offered by the Internet as well as local networks have made video data publicly and relatively easy available, for example Internet video collections, TV shows archives, video-on-demand systems, personal video archives offered by many public Internet services, etc. Video data should be indexed, mainly using content-based indexing methods. Digital video is hierarchically structured. Video is composed of acts, sequences, scenes, shots, and finally of single frames. In the tests described in the paper a special software called the AVI – Automatic Video Indexer has been used to detect shots in tested videos. Then, the single, still frames from different time positions in the shots detected in ten TV sports news have been thoroughly examined for their usefulness in an automatic classification of TV sports news.

9.1 Introduction

Recent advances in visual retrieval systems allow the storage of a large amount of digital video data. Furthermore, the new possibilities offered by the Internet and local networks have made video data publicly available. However, without appropriate indexing and retrieval methods all these video data are hardly usable. Users are not satisfied with the video retrieval systems mainly based on the same procedures as textual retrieval systems. They want to query not only technical data of videos such as length, format, type of compression, etc. but also the content of video clips [1]. For example, a user analyzing TV news can ask for specific events presented in the news sequence. Therefore, the need for effective tools to manipulate the video

Kazimierz Choroś
Institute of Informatics, Wrocław University of Technology,
Wybrzeże Wyspiańskiego 27, 50-370 Wrocław, Poland
e-mail: kazimierz.choros@pwr.wroc.pl

E. Tkacz and A. Kapczynski (Eds.): Internet – Technical Development and Appl., AISC 64, pp. 73–80.
springerlink.com

content is significant. Content-based search of videos becomes a difficult but at the same time fascinating problem. Generally, temporal segmentation is the first step of a video indexing process. The next is scene detection and content-based scene selection. To make the process of scene detection sufficiently effective the evaluation of usefulness of frames selected from video shots is needed. These single, still frames will serve as the basis for content-based analysis for scene detection process.

The paper is organized as follows. The next section describes the temporal segmentation process leading to the partition of a given video into a set of meaningful and individually manageable segments. They can serve as basic units for indexing and retrieval. The third section briefly discusses a detection of the sequence of shots making a scene. Moreover, some recent related research works are cited. In the sections 4 the process of shot and scene classification is described and the experimental results for the tested TV sports news videos are reported. Section 5 presents the analysis of the usefulness of the frames from different time positions in the shot and from the first, middle and the last shot of a scene for the classification of sport video shots. The final conclusions and the future research work areas are discussed in the last 6th section.

9.2 Temporal Segmentation

A video clip is structured into a strict hierarchy and is composed of different structural units. The most general unit is an act. So, a film is composed of one or more acts. Then, acts include one or more sequences, sequences comprise one or more scenes, and finally, scenes are built out of camera shots. A shot is a basic unit. A shot is usually defined as a continuous video acquisition with the same camera, so, it is as a sequence of interrelated consecutive frames recorded contiguously and representing a continuous action in time or space. The length of shots is very important. It greatly affect a film. Shots with a longer duration make a scene seem more slower paced whereas shots with a shorter duration can make a scene seem dynamic and faster paced. The average shot length of a film is generally several seconds or more.

A shot change occurs when a video acquisition is done with another camera [2]. The cut is the simplest and the most frequent way to perform a change between two shots, and at the same time cuts are probably the easiest shot changes to be detected. Cut takes place when the last frame of the first video sequence is directly followed by the first frame of the second video sequence. But mainly due to the application of digital movie editing software shot changes become more and more complex and more and more attractive for movie audience.

The other basic shot changes are fades and dissolves. A dissolve is a transition where all the images inserted between the two video sequences contain pixels whose values are computed as linear combination of the final frame of the first video sequence and the initial frame of the second video sequence. Cross dissolve describes the cross fading of two scenes. Over a certain period of time (usually several frames or several seconds) the images of two scenes overlay, and then the current scene dissolves into a new one.

Cross dissolve in digital environment – contrarily to analogue systems – is relatively easily realized. Dissolve can be modeled as luminance value operations. It is performed according to the following mathematical formula

$$Y = \alpha Y^1 + (1 - \alpha)Y^2. \tag{9.1}$$

The luminance Y of the pixels of dissolved image is the sum of the luminance values Y^1 and Y^2 of the adequate pixels of two frames from two mixed shots. The parameter α decreases from one to zero. The number of its values determines the number of frames (duration) for a given dissolve effect [2, 3].

Fades are special cases of dissolve effects, where a black frame most frequently replaces the last frame of the first shot (fade in) or the first frame of the second shot (fade out).

It is worth to mention that these special shot changes, special effects are not used in professional film editing. But not only young inspired amateur directors love digital effects. They often include effects or transitions like, for example, a wipe. A wipe effect is obtained by progressively replacing the old image by the new one, using a spatial basis.

Many tests, experimental works have been undertaken, many papers have been written on temporal segmentation processes. The efficacy of many methods have been evaluated and it can be said that we can state we can effectively detect shots in digital videos. Also specially designed, our software called the AVI – Automatic Video Indexer [4] have achieved sufficient level of quality and reliability (Tab. 9.1) to be applied in practice for further research investigations.

Table 9.1 The best results of recall R and the best results of precision P of temporal segmentation methods received (but not necessarily simultaneously) for several categories of video when testing the effectiveness of the Automatic Video Indexer

Results with the best recall and the best precision	Pixel pair differences		Likelihood ratio method		Histogram differences		Twin threshold comparison	
	R	P	R	P	R	P	R	P
TV Talk-Show	1.00	1.00	1.00	0.98	1.00	1.00	1.00	0.89
Documentary Video	0.87	1.00	0.98	1.00	0.89	1.00	1.00	0.86
Animal Video	0.88	1.00	1.00	0.89	0.96	1.00	1.00	0.76
Adventure Video	1.00	0.80	1.00	0.76	0.92	1.00	0.97	0.75
POP Music Video	0.95	1.00	0.85	0.90	0.65	1.00	0.88	0.85

In the tests described in the paper the Automatic Video Indexer has been used to detect shots in tested movies. Then, the detected shots have been thoroughly examined for their usefulness in scene detection process.

9.3 Scene Detection and Classification

The next step of an automatic video indexing is a detection of the sequence of shots making a scene. A scene usually corresponds to some logical event in a video such as a sequence of shots making up a dialogue scene, or an action scene in a movie. A scene is a part of the action in a single location. A scene can be master type, introduction or flashback scene, dynamic or static, long or short. Each scene in a movie should be long enough to show the action, but at the same time should be short enough to keep the audience's attention. If scenes run on too long, viewers may be feel bored and may lose interest. But if a given scene goes by too fast, viewers may miss the point.

There are many recent investigations towards automatic recognition of a content of a video clip [1, 5, 6], many of the detection methods have been tested on sport videos [7]. In the field of an automatic video processing research, a sport videos summarization has become a popular application because of its popularity to users and its simplicity due to repeated patterns.

In [8] a two-level framework has been proposed to detect automatically goals in soccer video using audio/visual keywords. The first level extracts low-level features such as motion, color, texture, pitch, etc. to detect video segments boundaries and label segments as audio and visual keywords. Then, two Hidden Markov Models have been used to model the exciting break portions with and without goal event respectively. The proposed approach has been applied to the detection of goal event in six half matches of soccer videos (270 minutes, 14 goals) from FIFA 2002 and UEFA 2002 and achieve 90% precision and 100% recall, respectively.

Other experiments have been carried out with baseball [9] as well as tennis videos [10]. Furthermore, many experiments have been also performed on sport classification [11, 12, 13].

9.4 Classification of Shots from TV Sports News

In our tests ten TV sports news videos have been used. The Automatic Video Indexer detected 333 shots. Then, three still frames of every shot have been shown to testers whose task was to tell us what sport discipline a frame comes from. The frames have been selected from three time positions in a shot: 10% from the beginning, 50%, and finally 90% from the beginning of a shot. Thus, the usefulness of the frames from three a priori chosen time positions in a shot has been examined.

The testers not only identified the sport discipline but also estimated the subjective certainty of their decisions on a scale of zero to ten. Table 9.2 presents the number of shots of every sport discipline and a subjective level of tester certainty of the identification decision.

The testers are not specialist of sports, i.e. they do not know the players of sport teams or their personal looks, they do not know the names and marks of sponsors of players or sport teams, they do not know the mottos or slogans of sport clubs, etc. They may decide only on such visual aspects of the presented frames like sport equipments or wears, place of a game, number of players, props of a game, etc.

The highest level of certainty has been observed in the case of ranking tables presented during TV sport news. It is not surprising. Ranking tables are easy for recognition and they are easily discriminated from other shots. On the other hand, the lowest level of certainty for news on the car racings is also not surprising, because during these TV news not only the bolides competing on racetracks are shown but also numerous, spectacular events and people accompanying the car racings.

Table 9.2 Usefulness of the frames from different time positions in the shots detected in ten TV sports news for the classification of sport video shots

Sport discipline	Number of shots	Average certainty [0-10]	Position of the key frame in a shot		
			10%	50%	90%
			Certainty level [0-10]		
1. Alpinism	2	5.00	5.00	5.00	5.00
2. Basketball	24	5.64	5.75	5.42	5.75
3. Closing credits	4	5.83	5.00	7.50	5.00
4. Car racing	24	3.61	2.33	3.96	4.54
5. Commentary	63	5.34	5.40	5.56	5.08
6. Golf	5	5.07	5.60	5.60	4.00
7. Opening credits	4	7.00	2.50	9.75	8.75
8. Preview	35	4.58	3.31	5.23	5.20
9. Ranking table	7	9.10	7.57	10.00	9.71
10. Ski jumping	11	6.24	5.82	6.45	6.45
11. Soccer	143	5.28	5.18	5.34	5.34
12. Speedway	11	5.64	5.64	5.64	5.64
	Averages:	5.29	4.93	6.29	5.87

The results presented in Table 9.2 may lead to a conclusion that the frames from the beginning of a shot are less useful for a sport discipline identification. The most useful are the frames from the middle part of a shot, a little bit less those from the end. The observation of the shots explains the reasons of such a situation. At the beginning of shots a player is usually close to the camera and then after zooming the court view dominates on the screen.

9.5 Best Frame Position in a Scene

Knowing that a scene is a set of shots it would be interesting to find which frame from which shot of a scene is the best for a classification of a whole scene, in our experiments which frame is the most adequate for a sport discipline identification. 333 shots detected in the tested ten TV sports news videos belong to 26 scenes which are composed of minimum two shots and, furthermore, there are 76 single shot scenes. There are 21 scenes which have at least five shots (Tab. 9.3). These 21 scenes were analyzed.

Whereas, Tables 9.4-9.6 present the evaluation of the usefulness of the frames taken from the first, middle, and finally the last shot of a scene. The results obtained are interesting.

Table 9.3 Characteristic of the scenes from ten TV sport news videos

Sport discipline	Number of scenes (min 2 shots)	Average length of scenes [shots]	Average certainty [0-10]	Number of scenes (min 5 shots)	Average length of scenes [shots]	Average certainty [0-10]
1. Alpinism	1	2.0	5.00	0		
2. Basketball	4	6.0	5.54	3	7.3	5.91
3. Closing credits	0			0		
4. Commentary	2	15.5	0.97	2	15.5	0.97
5. Car racing	1	24.0	3.61	1	24.0	3.61
6. Golf	1	5.0	5.07	1	5.0	5.07
7. Opening credits	0			0		
8. Preview	3	2.3	4.57	0		
9. Ranking table	0			0		
10. Ski jumping	1	11.0	6.24	1	11.0	6.24
11. Soccer	11	13.1	5.32	11	13.1	5.32
12. Speedway	2	5.5	5.64	2	5.5	5.64
Averages:		9.4	4.66		11.6	4.68

Let us remember that the best frame position in a shot is a middle part of a shot. But when analyzing the shots of a given scene the best strategy is to process the first shot of a scene, so, the beginning of a scene. The average certainty of the tester decision has significantly diminished, from 6.84 for the middle frame of the first shot to 2.81 for the final frame of the last shot in a scene. So, the best frame is a frame from the middle part of the first shot in a scene.

9.6 Final Conclusions and Further Studies

A video clip is structured into a strict hierarchy and is composed of several acts. Then, acts include one or more sequences, sequences comprise one or more scenes, and finally, scenes are built out of camera shots. The temporal segmentation process leads to the partition of a given video into a set of meaningful and individually manageable segments, which then can serve as basic units for indexing. There are many sufficiently effective segmentation techniques which are able to select shots by detecting cuts and cross dissolve effects in a movie clip.

An identification process of a sport discipline for series of shots detected in TV sport news has been discussed. The most simple and not time consuming process of

Table 9.4 Usefulness of the frames from the first shot of a scene for the classification of sport video shots

Sport discipline	Number of scenes	Position of the key frame in a shot			Average certainty
		10%	50%	90%	[0-10]
		Certainty level [0-10]			
Basketball	3	6.67	6.67	6.67	6.67
Car racing	1	0.00	8.00	10.00	6.00
Commentary	2	0.00	0.00	0.00	0.00
Golf	1	8.00	8.00	8.00	8.00
Ski jumping	1	10.00	10.00	10.00	10.00
Soccer	11	7.45	8.18	8.18	7.94
Speedway	2	5.00	5.00	5.00	5.00
Averages:		5.30	6.55	6.84	6.23

Table 9.5 Usefulness of the frames from the middle shot of a scene for the classification of sport video shots

Sport discipline	Number of scenes	Position of the key frame in a shot			Average certainty
		10%	50%	90%	[0-10]
		Certainty level [0-10]			
Basketball	3	7.50	7.50	7.50	7.50
Car racing	1	2.00	2.00	2.00	2.00
Commentary	2	5.00	5.00	5.00	5.00
Golf	1	0.00	0.00	0.00	0.00
Ski jumping	1	3.00	10.00	10.00	7.67
Soccer	11	4.64	4.73	4.73	4.70
Speedway	2	5.00	5.00	5.00	5.00
Averages:		3.88	4.89	4.89	4.55

Table 9.6 Usefulness of the frames from the last shot of a scene for the classification of sport video shots

Sport discipline	Number of scenes	Position of the key frame in a shot			Average certainty
		10%	50%	90%	[0-10]
		Certainty level [0-10]			
Basketball	3	5.00	7.33	7.33	6.56
Car racing	1	0.00	0.00	0.00	0.00
Commentary	2	0.00	0.00	0.00	0.00
Golf	1	10.00	10.00	2.00	7.33
Ski jumping	1	0.00	0.00	0.00	0.00
Soccer	11	5.27	5.36	5.36	5.33
Speedway	2	5.00	5.00	5.00	5.00
Averages:		3.61	3.96	2.81	3.46

identification is the analysis of a single frame taken from a shot. The problem arises which frame is the most adequate for recognition of a sport discipline. The tests have indicated that the most adequate frame is not the frame from the beginning of the shot, what is frequently practiced, but rather from the middle part or from the end of a shot. Further investigations lead us to the conclusion that if the automatic process of the identification of a sport discipline is based on a single, still frame, such a frame should be chosen from the middle part of the first shot in a scene.

In the further research we will test the influence of the length of a shot and of a scene on the usefulness of a single frame for an identification process, the most frequent length of a shot and the most frequent number of shots in a scene for a given sport discipline, and finally we want to extend the functionality of the Automatic Video Indexer using the automatic extraction of video features.

References

[1] Lew, M.S., Sebe, N., Djeraba, C., Jain, R.: Content-based multimedia information retrieval: State of the art and challenges. ACM Transactions on Multimedia Computing, Communications, and Applications, 1–19 (2006)

[2] Choroś, K.: Digital video segmentation techniques for indexing and retrieval on the Web. In: Advanced Problems of Internet Technologies. Academy of Business, Dabrowa Górnicza, pp. 7–21 (2008)

[3] Choroś, K.: Cross dissolve detection in a temporal segmentation process for digital video indexing and retrieval. In: Information and Computer Systems – Concepts, Tools and Applications. Oficyna Wydawnicza Politechniki Wrocławskiej, Wrocław, pp. 189–200 (2008)

[4] Choroś, K., Gonet, M.: Effectiveness of video segmentation techniques for different categories of videos. In: New Trends in Multimedia and Network Information Systems, pp. 34–45. IOS Press, Amsterdam (2008)

[5] Chena, L.-H., Laib, Y.-C., Liaoc, H.-Y.M.: Movie scene segmentation using background information. Pattern Recognition, 1056–1065 (2008)

[6] Money, A.G., Agius, H.: Video summarisation: a conceptual framework and survey of the state of the art. Journal of Visual Communication and Image Representation, 121–143 (2008)

[7] Zhong, D., Chang, S.-F.: Real-time view recognition and event detection for sports video. Journal of Visual Communication and Image Representation, 330–347 (2004)

[8] Kang, Y.-L., Lim, J.-H., Kankanhalli, M.S., Xu, C., Tian, Q.: Goal detection in soccer video using audio/visual. In: Proceedings of the ICIP, pp. 1629–1632 (2004)

[9] Lien, C.-C., Chiang, C.-L., Lee, C.-H.: Scene-based event detection for baseball videos. Journal of Visual Communication and Image Representation, 1–14 (2007)

[10] Delakis, M., Gravier, G., Gros, P.: Audiovisual integration with Segment Models for tennis video parsing. Computer Vision and Image Understanding, 142–154 (2008)

[11] Messer, K., Christmas, W., Kittler, J.: Automatic sports classification. In: Proceedings of the 16th International Conference on Pattern Recognition, pp. 1005–1008 (2002)

[12] Wang, D.-H., Tian, Q., Gao, S., Sung, W.-K.: News sports video shot classification with sports play field and motion features. In: ICIP 2004 International Conference on Image Processing, pp. 2247–2250 (2004)

[13] Ling-Yu, D., Min, X., Qi, T., Chang-Sheng, X., Jin, J.S.: A unified framework for semantic shot classification in sports video. IEEE Transactions on Multimedia, 1066–1083 (2005)

Chapter 10
E-Learning Database Course with Usage of Interactive Database Querying

Katarzyna Harezlak and Aleksandra Werner

Abstract. The problem of database issues teaching with usage of Internet tools was discussed in the paper. For this purpose the database knowledge and groups of course trainees were categorized. It has influenced the construction of a database e-learning course for which the module structure has been proposed. The designed course was implemented to the MOODLE platform. Various mechanisms for database knowledge presentation and practicing were used. This mechanisms were extended of a new activity making a database interactive querying possible. The process of the MOODLE platform extending and benefits of its usage were also presented.

10.1 Introduction

Universal access to the Internet and growing diversity of tools enabling internet usage for teaching purposes make many education institutions enrich their teaching offers of e-learning courses In particular Universities should not be and are not indifferent to such mechanisms. In many cases e-learning platform is used for materials exchanging between lecturers and students, but possibilities of these tools are much more sophisticated.

The purpose of the paper is to present a project and a simple implementation to MOODLE platform of database e-learning course.

Nowadays, it is difficult to come across a company that could function without efficient information management. Because of this, problems of databases designing, creating and management become the center of interests. Preparing a good database course project requires defining groups of course trainees appropriately to their level of computer science knowledge. There were four such groups considered

Katarzyna Harezlak · Aleksandra Werner
Silesian Technical University, Gliwice, Poland
e-mail: katarzyna.harezlak@polsl.pl,
e-mail: aleksandra.werner@polsl.pl

E. Tkacz and A. Kapczynski (Eds.): Internet – Technical Development and Appl., AISC 64, pp. 81–89.
springerlink.com

in the research: pupils of secondary schools, students of computer science and other divisions and post diploma persons. This analysis of database issues [2, 8] leads to the conclusion that database knowledge should be categorized, as various database problems can be presented in different ways to various groups of interested users. For this purpose database theory was divided to groups of issues as follows:

1. relational data model basics and its comparison with others,
2. relational data model design methods:

   a. normalization process,
   b. entity relational diagram,

3. creation of database and database objects,
4. SQL - data query language,
5. extended SQL programming,
6. database server management,
7. basics of database application programming.

Necessity of database knowledge categorization has influenced the construction of a database e-learning course. The characteristic features of the proposed structure are its modularity and ability of extending its content, which guarantees needed flexibility. The course-work was designed as separate modules, in which various resources and activities, suitable for given group and therefore ensuring influence on student activity, motivation and commitment development are used. Another problem in database teaching is a sequence in which database issues should be presented. Trainees, regardless of which group they belong to, should start the study from getting to know relational data model basics. Next they ought to acquire ability of simple SQL query defining. Subsequent steps will depend on a given person or group needs. Sample flow of possibility of the database knowledge obtaining was presented in the Figure 10.1.

10.2 General Structure of a Module

Before we get down to fill a module with a content it is necessary to consider how its structure is supposed to look like to make database issues as accessible as possible. It should be simple, universal and uniform to allow trainee being inside the module to quickly find a way to the parts of information that he is interested in. Simplicity of usage ought to concern organizational matters as well. Therefore the general structure of the module was proposed. It contains two sections.

Information section consists of following elements: the scope, skills a trainee achieves after the module completion, ways of the module passing, ways of communication with a lecturer and other trainees, hardware and software needed to the module realization. While teaching section includes: theoretical knowledge, set of exercises preserving acquirement, tasks required for the module passing.

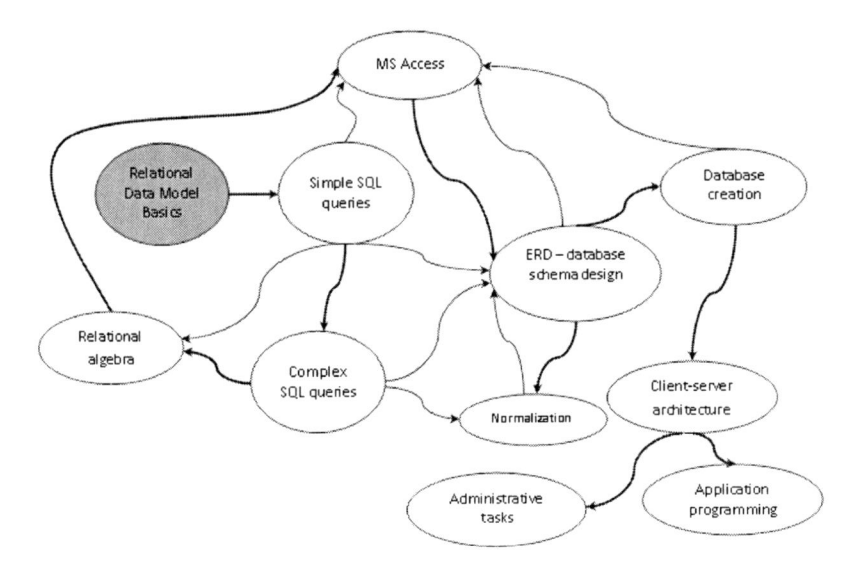

Fig. 10.1 Sample flow of the database knowledge obtaining

For implementation of the described earlier database course project the MOO-DLE e-learning platform was chosen in which author of the module can use two types of mechanisms. There are *resources* allowing the knowledge presentation and *activities* giving trainees possibility to test and to resolve given problems in practice.

Using such mechanisms, the designed course was divided into seven modules according to described in chapter 10.1 assumptions. Each module is composed of topics presenting defined part of issues characteristic for given module. In the beginning of each module links to module authors' emails and basic organizational information were placed as well (left side of Fig. 10.2).

Every topic was built in similar way. It contains a part devoted to knowledge presentation, in which HTML pages with conjunction with links to files were mainly used. Theoretical content was enriched by sets of examples and tasks with usage of lections, quizzes and topics for discussion via forums. Examples of these elements are presented on the chosen module concentrating on basic query language to relational databases in the Fig. 10.2.

Mechanisms existing on MOODLE platform allows diversifying ways of knowledge passing on, however in this case, the most interesting seems to be a possibility of interactive database query.

10.3 Description of SQL Learning Problem

In SQL language distance learning, the main challenge that the course-designer faces, is to guarantee the students an access to the database servers - freeware/

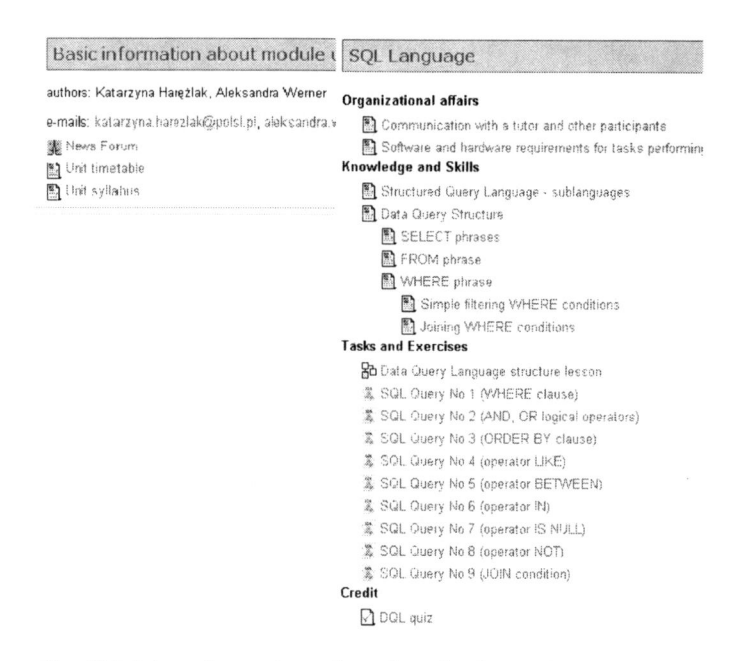

Fig. 10.2 Information section and sample topic of a sample module

shareware licensing or from MSDNAA program - that is later connected with individual installation process by course participants.

As far as database servers installation processes studying can constitute the separate topic of a course, from the didactic coherence point of view, it is not advisory to do that by the way of SQL queries training, because the participants additionally could face the problem of configuring the system installed on their computers properly. Taking the diversity of computer hardware and software solutions (for ex. different operating systems) owned by trainees into consideration, even the description of installation flow occurs to be not an easy and complex task. It usually causes introducing the assumptions, regarding students hardware equipment, by a course designer. It is worth emphasizing, that imposing too restrictive limitations is always connected with reducing the number of students interested in participating in e-learning course, that is undoubtedly an unfavorable phenomenon.

In order to avoid e-students' disinterest, relational database e-courses are usually designed in the way that allows the learners to train acquired skills on only one, free database system (for ex. PostgreSQL) chosen by a teacher. At the same time, system-depending SQL language variants are taught only theoretically (by the use of passive MOODLE resources). Obviously, such an approach results in the absence of the possibility of interactive Transact SQL language (MS SQL Server system) or PL/SQL (Oracle system) study. Thus, education process is incomplete and inefficient or requires using blended-learning formula. In case of a situation that students install prescribed database system on their personal computers, it must be secured that they will use the uniform database structures and uniform objects' names.

Consequently, every database schema changing, caused by - for example - adding new table necessity, must be immediately propagate to individual students' databases. It can arise the risk of mistaken query execution on out-of-date database, and - what is more - getting the wrong result set by students.

Specified inconveniences arising in SQL language on-line teaching point indicate the need of using solutions, which will maximally reduce the opportunity of potential students' mistakes occurrence and will protect against unexpected difficulties on this education stage as well.

These mechanisms - at the same time - mostly should ensure effective training people, who belong to different groups of course recipients and realize didactic programs with different degree of difficulty, not necessarily being able to cope with database system installation or administration.

After examining resources and activities offered by MOODLE platform and analyzing the rules of its operation authors found, that the most desirable platform functionality - solving touching problems, would be availability to interactive query execution, performed on database and managed by a subject teacher.

Besides, authors remark, that there is a possibility of realizing mentioned concept just in MOODLE environment, slightly revising its code. It can be reached through specific MOODLE properties - such as [3]:

- Open Source distribution, offering practical accessibility to a needed software's source,
- free software basing, like Apache, PHP, MySQL,
- modular organization allows to add new functionalities in fairly easy way,
- cooperation with a database server (here: MySQL), storing the majority of data needed to system running.

10.3.1 Developed Solution

Taking the referred problems into consideration, authors propose a solution that lies in the new interactive module creation, named SQL, accessible from the *Add an activity...* drop down menu connected with a topic of the course (Fig. 10.3). Once the SQL plug-in is available, instructors can add SQL assignments to their course.

During the implementation authors used - as a sample - a code already existing in the system, which was modified (adopted) for a new module functional needs [6]. According to the fact that MOODLE is written in PHP and modules are similar in structure, in practice authors had to rewrite parts of the certain *.php* scripts (but weren't forced to compose whole component from the beginning). In order to achieve this, the obligation installations were run as follows:

1. integrated software-database environment WAMPSERVER with MySQL (recommended) database,
2. Content Management System MOODLE,

and after analyzing the content of moodle folder, it turned out that only folders *mod* and *lang* should be modified.

Concurrently it was observed that maintaining uniform naming is very important (in *mod* folder, every subfolder contains code for a module of corresponding name - new added by authors activity *SQL* as well - Fig. 10.3).

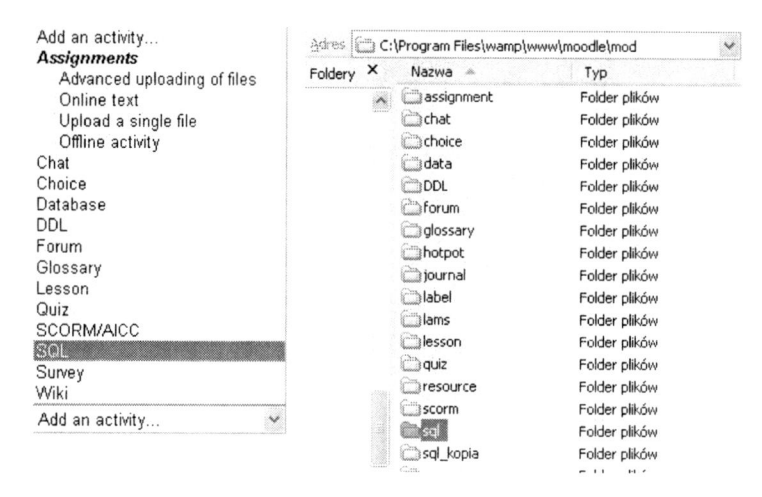

Fig. 10.3 List of activities available in MOODLE platform and list of MOODLE platform PHP-writing modules, enclosed in a separate folders

So, the naming of new module comes from two places [4].

1. First is the directory name with the template files - ie. */mod/modulename*. Many links refer to this directory.
2. Second are the string files in the *lang* folder. There are a few strings that are used by Moodle to identify new module on the admin page and when adding a resource.

Activity modules reside in the *mod* directory. Each module is in a separate subdirectory and consists of the following mandatory elements (plus extra scripts unique to each module)[5]:

The major modifications of SQL module were made in files [1]:

- **mod_form.php** - contains the definitions of new added module parameters (i.e. definitions of fields of displayed form) and is responsible for a contents of page, seen by a teacher during activity editing,
- **view.php** - contains a code realizing SQL queries executing on chosen database and is responsible for contents of page, seen by student during updating the assignment connected with the activity. Inside this file, every database server has its separate block of code, which general structure is listed in the listing 1.

```
create a connection variable;
if (connection established){
      create the variable that holds SQL statement;
      parse the statement in the context of a database connection;
      execute the statement;
      while (there are rows in the result set){
           get the number of columns used in SQL statement;
           for (\$ i is less than number of columns){
             return column value for fetched row;
           }
        }
    }
else
      print an error message and exit the program;
free up the resources used to perform the query;
close the database connection;
```

Listing 1. Pseudo-code of SQL query execution implementation

Additionally, elements appearing on forms of both files (for ex. labels and) have their own language equivalents written in *sql.php* files of *.../MOODLE/Lang/en_utf8/* and *.../MOODLE/Lang/pl_utf8/* folders:. Fragment of *.../MOODLE/Lang/en_utf8/sql.php* file is presented on listing 2.

```php
<?php
\$ string['modulename'] = 'SQL';
\$ string['modulenameplural'] = 'SQL';
\$ string['db_host'] = 'Database host';
\$ string['db_name'] = 'Database name';
\$ string['enter_sql'] = 'Enter SQL query';
\$ string['result_set'] = 'Query result';
...
?>
```

Listing 2. Sample UTF8 string language manipulation

To the needs of *view.php* file revision, the possibility of establishing the connection with remote database servers was reflected on. Thus, database servers extensions, allowing developer to access the databases, were studied. Finally, connection with: IBM DB2, MS SQL Server, MySQL and Oracle 11g was provided in *view.php* file. Databases can be accessed through a local or remote database server.

The developed SQL plugin gives users of the MOODLE Learning System the ability to on-line SQL queries execution. These queries are running on a denoted database, that resides on university database server.

The choice of a database - it means: type of database engine, its host and database user name/password - takes place during the edition of describing activity (Fig. 10.4). Although the default database server is MySQL - the database, which main task is storing the data of MOODLE platform - it is also possible to query another database systems, such as: Oracle, MS SQL Server or IBM DB2 (Fig. 10.4), which are used during respondent stationary-mode database subject study.

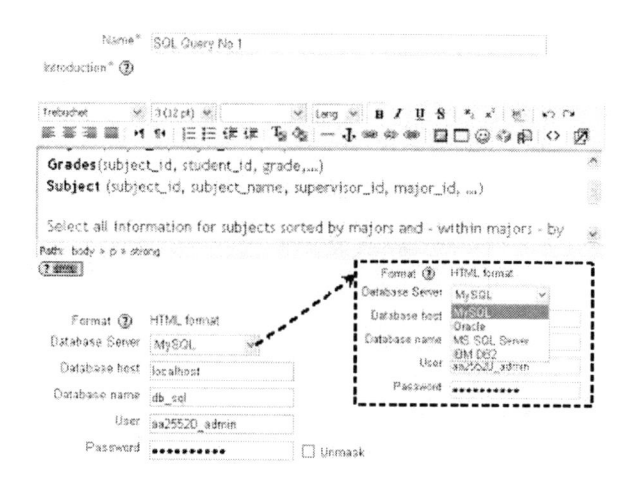

Fig. 10.4 View of SQL activity edition window

10.4 Conclusion

The subject of SQL query distance learning, by the use of MOODLE platform, is taken up in the article. In connection with this issue, subject knowledge and course participants target groups were categorized. The structure of such e-course was suggested and sample MOODLE-course realization was made. One of the most interesting developed solution, improving the e-learning database platform, was MOODLE resources extension of a remote database server communication possibility. New activity enable students to interactively practice theoretical information, published on a platform. The main advantages, gaining from proposed MOODLE activities extension, are:

1. opportunity of introducing the students to SQL language standards, used by particular database systems,
2. the possibility of making potential modifications of database schema entries only in one place,
3. fast and trouble-free database servers updating (for example in the case of software migration or environment modernization),
4. client-space security control.

Successfully finished tests of added SQL activity causes making an attempt to another module creation, which would enable students to perform some DML (such as: INSERT, UPDATE, DELETE) and DDL statements. Because of meeting a difficulties (for ex. absence of appropriate functions in extensions of all used database servers), authors decided to implement specified activity by the use of stored procedures mechanism, but the researches still last.

References

1. Community site of Moodle platform (2009), `http://moodle.org/`
2. Date, C.J.: An Introduction to Database Systems. Addison-Wesley, Reading (2000)
3. Dougiamas, M.: MOODLE - Developers Manual (2006),
   `http://cvs.moodle.org/lang/sm_utf8/docs/developer.html?view=co`
4. Kalis, P.: How to install new module to MOODLE (2008), `http://tamingmoodle.`
   `blogspot.com/2008/01/moodle-is-being-said-to-be-great-at.html`
5. MOODLE - Activity Modules (2009),
   `http://docs.moodle.org/en/Development:Modules`
6. New Module Template Code (2009),
   `http://download.moodle.org/plugins/mod/`
7. PHP Manual - Database Extensions, PHP Documentation Group (2009),
   `http://www.php.net/manual/en/refs.database.php`
8. Ullman, J.D., Widom, J.: Podstawowy wyklad z systemow baz danych. WNT, Warszawa (2000)

Chapter 11
On Applications of Wireless Sensor Networks

Adam Czubak and Jakub Wojtanowski

Abstract. Wireless Sensor Networks (WSN) are ad-hoc networks in which small independent sensor nodes have limited energy, computational resources and wireless communication capabilities. Recently, both academia and industry have shown great interest in the area of Wireless Sensor Networks. This paper focuses on the practical applications in commerce and feasible future employment of WSNs. Continued advances of wireless communication technologies have led to the following fields of applications: habitat and environmental monitoring, security of buildings and property, oil and gas installations, mines and tunnels, emergency medical care, military applications. In the near future WSNs will certainly enter our homes and offices changing the way we monitor our nearest surrounding.

11.1 Introduction

A Wireless Sensor Network is a group of small devices (*nodes*) with an integrated energy source and a capability for wireless communication. It is an autonomous micro-sensor network which purpose is to detect, acquire and report particular events across a remote area. These can be deployed in areas as diverse as a battlefields or an emergency rooms. The technology utilizes communication channels in the ISM band defined by ITU-R (usually 900MHz or 2.4MHz) and therefore, unlike other wireless technologies, does not require clear line of sight between transmitter and a receiver. Wireless Sensor Networks are designed to handle very low data throughput (even as low as few bits/day), exchanging lower throughput and higher message latency for longer battery life, lower cost, and self-organization[1]. Employees, who earlier

Adam Czubak · Jakub Wojtanowski
Opole University, Institute of Mathematics and Computer Science,
ul. Oleska 48, 45-052 Opole, Poland
e-mail: {adam.czubak,jakub.wojtanowski}@math.uni.opole.pl
http://www.math.uni.opole.pl

E. Tkacz and A. Kapczynski (Eds.): Internet – Technical Development and Appl., AISC 64, pp. 91–99.
springerlink.com

deployed traditional data networks install the sensors just like a wired device but neglect the cables. WSNs are logically plugged into the existing network and there is no distinction between the data which originated at the wireless sensors and the data that was received from a conventional wired device. The nodes operate like wired devices and are also commissioned like such.

11.2 Sensor Networks from a Practical Point of View

The advantages of Wireless Sensor Networks are as follows:

Ease of deployment and removal   The placement of the sensors may be random or deliberate. It may be as easy as scattering a handful of coins on the ground or dropping them from a helicopter onto a battlefield. These have the ability to establish connections to each other, organize and send the information across the network to the gathering device (*sink*) regardless of their location. These devices are small, require no wiring and messy construction work, an employee responsible for their allocation needs neither training nor complicated and expensive tools. When the WSN is no longer required it is easy to remove and reuse in a different location. This characteristic makes the solution very cost-effective and thus widely available.

Sensing   equipment with which a node may be equipped is practically unlimited, ranges from pressure sensors, temperature, humidity, flow, position, velocity, angular velocity, acceleration, strain, force, torque, slip, vibration, contact, proximity, distance, motion, to biochemical agents and liquid sensors.

Size   varies depending on the manufacturer and applied technology but nowadays a single DN2600[1] 12 x 12 x 1.58mm chip is a compilation of a microprocessor, IEEE 802.15.4-compliant radio and integrated power amplifier for 10 dBm operation. The solution only lacks the battery and an antenna. So the size of the whole device is actually as small as the size of the installed battery or the length of the antenna, whatever is larger.

Locality   means within the range of the sensor network. Current protocols deal reliably with up to 500 nodes. Their communication range is up to 50m in radius. That means that a regular sensor network utilising well known standards and hardware solutions without any tweaks or upgrades may span over 25 kilometres in a straight line.

If we take a closer look at the buildings we use today we notice, that these are very sophisticated structures. Walls are filled with conduits for sensors: fire sensors, temperature sensors, cameras, light sensors. The purpose of these systems is to make our surrounding safer, comfortable, dynamic and energy-efficient. All these sensors are wired in a network. The sensors and the computing power is getting cheaper but the cost of conduits is not and the install cost of conduits, labour and material for

---

[1] DN2600 is a 2.4GHz Mote-on-Chip™ designed by Dust Networks. It is a part of a solution called SMARTMESH SA-600™ http://www.dustnetworks.com/

the circuits remain uniform or increase over time. Another aspect worth mentioning is that the conduits required for classic sensors installed along the walls are bad-looking and troublesome. WSNs have the potential to reduce the overall costs of such installations.

Therefore the industry's enthusiasm for the new technology grew and an indus-trial ZigBee Alliance was formed and a standard called ZigBee was introduced in 2004. In the first(physical) and second(MAC) layer it was based on IEEE 802.15.4 radio standard. Due to its inefficiency in 2006 a second ZigBee standard was intro-duced which was incompatible with the previous one and in 2007 a third one called ZigBee Pro emerged. But very little devices were build using these solutions and there was no real grow in market. Nevertheless, the standardization of 802.15.4 and ZigBee mark the milestone in development of Wireless Sensor Networks introduc-ing common platforms to design from scratch or integrate platforms of WSNs. It introduced certified devices and proved that these are actually low cost installations. A survey published in 2005 conveyed, that the deployment of WSNs is hindered for the following reasons (see Fig. 11.1).

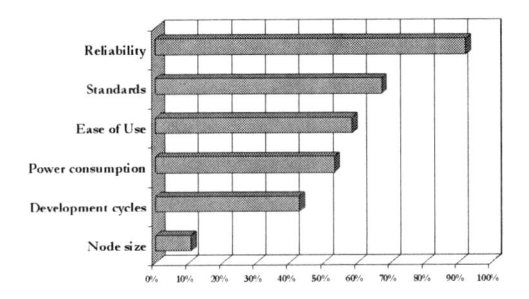

Fig. 11.1 Source: OnWorld'05. *WSN: Growing Markets, Accelerating Demand*

The industry applications required reliability for the data gathered with sensor networks, that ZigBee did not provide given the challenges of the ISM RF environ-ment (e.q. frequent interferences). This situation gave rise to other companies and organisations like Dust Networks, HART Communication Foundation, ISA et. al. Their research departments either developed their own solutions or worked together on the WSN issues.

Contemporary standards provide the following basic characteristics[2]:

- Time to send and receive a frame with data equals 9ms. This value corresponds to a process of booting up a transmitter, sending the packet, receiver startup time, lo-cal CRC check, ACK calculation and transmission time of the acknowledgement. It it the total time required for transmission of a single chunk of data between two nodes. So it allows over a 100 complete and verified transmissions per second.

---

[2] Sample data for Dust Networks solutions utilising DN2600 chips and Time Synchronized Mesh Protocol (TSMP) (http://www.dustnetworks.com).

- The above process costs $20\mu A$ of energy. It is the lowest power consumption for delivering packet available nowadays in WSNs while retaining 99.9% reliability. This low cost communication results in a lifetime of around 10 years (per node) in contemporary applications when using two standard AA batteries as an incorporated energy source.

11.3 Technological Aspects

From a technological point of view the primary problem in WSNs is the limited power supply that a node possesses. The challenge here is to use it as rarely as possible saving as much energy as possible. Battery is drained by:

- the sensing equipment which collects the data,
- the microprocessor when analysing the collected information and making decisions whether to send it or not and towards which node,
- the communication equipment i.e. radio transceiver when sending, receiving or relaying the information.

Of the three above the cost of communication is the highest[2]. The energy $(E_{i,j})$ required for transmitting a message of size m from node i to node j over a distance d equals $E_{i,j} = c + md^2$ where c is a constant amount of power required for booting up the transceiver. The model seams fairly simple but in the real world it turns out, that the distance d is squared only in perfect vacuum. So the exponent is always higher than 2 depending on the surrounding.

Therefore it is of utter importance that the amount of data sent across the WSN is as small as possible. There are several ways to tackle the finite battery problem:

- Data aggregation and data fusion[3, 4]. Aggregation and compression of data allows to reduce communication cost
- Recovery effect. The lifetime of the battery depends on the characteristics of the discharge process. If the battery is being discharged only for short time intervals followed by idle periods improvements in energy level is possible. During the idle periods the battery has the capacity to partially recover the capacity lost in the preceding transmission[5, 6]
- Energy harvesting. Energetic sustainability[7] can be achieved by embedding recharging devices like solar panels in the node.
- Standby scheduling[8]. Allowing the sensors to turn off almost completely while not being in use is a great advantage but requires precise time synchronization between nodes.
- Intermediate power-attached nodes (mediation devices) used as relays or buffers for the data traversing the network.
- High-level data descriptors. So called meta-data to eliminate the transmission of duplicate and redundant data throughout the network.

Another strategy would be to tweak the routing process and send data only over short, not so energy-consuming distances. This policy, although locally for a single node reasonable, turns out to be very expensive if we consider the overall cost of sending the data across the network by a large number of short jumps instead of a single, longer one. So it is important that protocols used in WSN work in a manner that both saves energy of the individual node and manages summarized energy of the system equally and fairly. Another issue is, that any node which loses power drops out of the communications network and may end up partitioning the network.

It is worth mentioning, that fully connected networks suffer from problems of NP-complexity[9]. As new nodes appear in the network, the amount of links increases exponentially. Therefore, for large networks, the routing problem is computationally intractable even with the availability of large amounts of computing power.

11.4 Object, Habitat and Environmental Monitoring

Sensors provide scalable solutions for security and surveillance companies. This kind of applications may be set up in different environments such as private properties, urban areas, etc. Constant cooperation among networked nodes increase the security of the concerned environment without human intervention. The system should respond to the changes of the environment as quickly as possible contacting the responsible security officer.

The possibility of detection of relevant quantities of poisonous gases in mines or tunnels, measuring methane or other gas levels in mines and tunnels could avoid disasters and save countless lives. For example, an operator could remotely supervise different physical phenomena in the mine from his computer and thus provide safe air/oxygen for the miners underground by monitoring the level of methane and other noxious gases. Such environment monitoring system called SASA[10] was introduced and a prototype was deployed in a coal mine.

The detection of seismic activities[11] with help of sensitive detectors, connected wirelessly to a system for recording information, gather the wave data in areas where earthquakes occur. The records of detected waves help to find the epicentre of the earthquake in a very short time. This allows to inform instantly earthquake safety teams and organize necessary help.

Grape Networks, Inc. currently ships commercial sensor networks for vineyards. Each node can sense a couple of parameters such as temperature, moisture or light. The system gathers enormous sums of data. With this solution crops can be monitored on a much finer level than even today's agriculture techniques and span over larger areas.

An Intelligent Car Park Management System based on Wireless Sensor Networks was recently introduced. A prototype was build called Smart Parking (SPARK) Management. It provides advanced features like remote parking monitoring, automated guidance and parking reservation mechanism. It allows the minimization of time usually spent on finding a parking spot which is a common issue in larger cities.

11.5 Building Automation

When it comes to building automation and a term *smart building* sensor networks are a perfect solution. Smart building represent the next step in evolution of this industry. Such buildings rely on the sensory data gathered throughout the region. This makes the building more interactive and more *aware*. Sensors help to improve the energy management by building enhanced energy usage models. It resulted in solutions which measure, detect, estimate, and predict the variable environmental factors for energy expenditure like for example energy-efficient lighting control.

Areas such as security and surveillance have already been explored to some extent and WSN devices are available like the MICA based surveillance systems (MICAz by Crossbow Technology http://www.xbow.com). Admission control is also a field where tiny sensor provide high functionality at low cost[12].

The HVAC systems are very energy consuming. A set of sensors measuring the current temperature and humidity allow to build a system which reacts instantly and seamlessly when a variable changes in order to provide comfort. But when no person is in the building it would turn the system off saving energy costs. Such nodes and systems can be purchased at Siemens AG. Hotels and offices can benefit from wireless sensors and controllers. The savings can be significant because these sensors work 24/7 and can be easily adjusted with software controls instead of manual adjustments to thermostats, lights and appliances. The electrical usage may be reduced by 20-40%.

11.6 Oil and Gas Installations

WSN is a technology now deployed in key investment area across the whole oil and gas supply chain including pipelines, refineries, petrochemical plants, exploration and production.

The reason for that are growing costs of labour and the fact that oil and gas supplies become more difficult to find and exploit. Additionally, refineries and petrochemical plants are faced with stringent regulations and are operating at near full capacity. By providing secure and reliable two-way wireless communications, WSN enables automation and control solutions that are not feasible with wired systems to improve the whole production process. The solutions are widely available now. The most common is Emerson Smart Wireless (Emerson Electric Co. http://www.emersonprocess.com/SmartWireless) which provides wireless sensors measuring pressure, temperature, level, flow, vibration, discrete switches, or pH. The reduction of costs is tremendous and these solutions became a standard for new installation. In fact, oil and gas industry was the first to deploy WSNs successfully on a large scale.

11.7 Medical Care

Wireless sensor network research is being performed to address medical applications. But in medical application node size matters. When a patient has wireless sensors attached to their body these must be as little intimidating and as light as possible. These should not impede in any way the patients regular activities.

It is important to distinguish pre-hospital and in-hospital emergency care. The fact that both can be realized using the same sensor set makes the care a real 24h service. It therefore allows disaster response with a minimal possible reaction time and more efficient patient rehabilitation.

The AID-N project (`http://www.aid-n.org`) at Johns Hopkins Applied Physics Laboratory proposes a device called miTag. miTag monitors a patient's location, heart rate, and oxygen saturation and relays this information over a WSN through a sink to monitoring stations. Thousands of patients can be monitored simultaneously over the miTags' resilient wireless network. These devices provide services like: patient monitoring, tracking and documentation regardless of the patients location.

The Mercury Project (`http://fiji.eecs.harvard.edu/Mercury`) at Harvard University created a wearable sensor network platform for high-fidelity motion analysis. A Mercury network consists of a number of wearable sensors called SHIMMERS each 4 x 3cm in size. Each sensor samples multiple channels of accelerometer, gyroscope, and/or physiological data and stores raw signals to local flash. Sensors also perform feature extraction on the raw signals, which may involve expensive on-board computation. Used mainly in treatment of Parkinson's disease at a local hospital.

Eldercare is a pre-hospital in-home health care. A system was introduced by Intel (`http://www.agingtech.org`) and actually made it possible for elders with Alzheimer's disease, degenerative disease of nerve cells and others with disabling conditions to continue to live at home.

Nowadays WSNs are also used for Exercise ECG, Signal-averaged ECG and a Holter monitor. The last one is especially suitable for WSNs since in this case the ECG recording is done over a period of 24 or more hours usually during patient's regular workday.

11.8 Military Applications

Due to the fact, that information on military applications of any new technology is not disclosed to general public there is very little to conclude. But there was an interest in using WSNs for perimeter security. One of US defence contractors (SAIC) developed a solution for this purpose. The sensors used were: magnetometer, seismometer for detecting footsteps or vehicles, passive infra-red, microphone and camera all integrated in a 10 x 10 x 5 cm box. The software system records the events in the log allowing for both real-time remote observation of the perimeter as well as later analysis.

PinPtr (http://www.isis.vanderbilt.edu), an acoustic system, takes advantage of sensor network technology to find the location from which a shot (e.g. with a sniper rifle) was taken. Instead of using a few expensive acoustic sensors, a low-cost ad-hoc acoustic sensor network measures both the blast and shock wave to accurately determine the location of the shooter and the trajectory of the bullet. The basic idea is simple: using the arrival times of the acoustic events at different sensor locations, the shooter position can be accurately calculated using the speed of sound and the location of the sensors. Precision with which the source of the shooter's location was found is 1m.

Other applications like battlefield surveillance via aerial deployment of sensors and body sensors for soldier condition analysis have been rumoured about for a long time but no official information was disclosed.

11.9 Conclusions

The applications for WSNs are many and varied, but typically involve some kind of monitoring, tracking, and controlling. Their presence in our surrounding with time will occur more often. The contemporary solutions are ready to use and reliable.

But besides the functional and technological aspects of WSNs there is also an ethical one. These sensor networks possess a great potential for abuse. Questions like: Who is controlling the system? What is the system collecting? Who is controlling the network? Who has access to it? Where and for how long will the data be stored? The devices are merciless, they collect information that is also not necessary or even unwanted and people may not feel comfortable around this technology.

Nevertheless it is clear that both the industry and academia acknowledged the new technology and also advances are imminent.

References

1. Callaway Jr., E.H.: WSN: Architectures and Protocols. CRC Press, Boca Raton (2003)
2. Tai, S., Benkoczi, R., Hassanein, H., Akl, S.: Implementation Guidance for Industrial-Level Security Systems Using Radio Frequency Alarm Links. In: IEEE International Conference Communications, pp. 3432–3437 (2006)
3. Sang, Y., Shen, H., Inoguchi, Y., Tan, Y., Xiong, N.: Secure Data Aggregation in Wireless Sensor Networks: A Survey. In: Proc. of PDCAT 2007, pp. 315–320 (2006)
4. Liang, B., Liu, Q.: A Data Fusion Approach for Power Saving in Wireless Sensor Networks. In: Proc. of IMSCCS 2006, pp. 582–586 (2006)
5. Chiasserinia, C.F., Rao, R.R.: Stochastic battery discharge in portable communication devices. In: IEEE Aerospace and Electronic Syst. Mag., pp. 41–45 (2000)
6. Chiasserini, C.F., Rao, R.R.: Routing protocols to maximize battery efficiency. In: Proc. of MILCOM 2000, pp. 496–500 (2000)

7. Lattanzi, E., Regini, E., Acquaviva, A., Bogliolo, A.: Energetic sustainability of routing algorithms for energy-harvesting wireless sensor networks. Computer Communications 30, 2976–2986 (2007)
8. Keshavarzian, A., Lee, H., Venkatraman, L.: Wakeup scheduling in wireless sensor networks. In: Proc. of the 7th ACM Int. Symp. on Mobile ad hoc networking and computing, pp. 322–333 (2006)
9. Garey, M.R., Johnson, D.S.: Computers and Intractability: a Guide to the Theory of NP-completeness. Freeman, San Francisco (1979)
10. Li, M., Liu, Y.: Underground coal mine monitoring with wireless sensor networks. ACM Trans. Sen. Netw., 1–29 (2009)
11. Horton, M., et al.: Deployment ready multimode micropower wireless sensor networks for intrusion detection, classification, and tracking. In: Proc. of C3I, pp. 290–295 (2002)
12. Swank, R.G.: Implementation Guidance for Industrial-Level Security Systems Using Radio Frequency Alarm Links, Westinghouse Hanford Company Technical Security Document WHC-SD-SEC-DGS-002 (1996)

Chapter 12
Gift Cards Authorization through GSM in a Distributed Trade Network – Case Study

Joanna Zukowska and Zdzislaw Sroczynski

Abstract. In today's rapidly changing economy it is vital to maintain appropriate relationships with clients. Any kind of loyalty schemes and gift cards systems, present in small and medium-sized trading networks, require the use of appropriate information technology. However, the cost and organizational factors and common lack of access to the Internet, causes the need to develop dedicated organizational and technical solutions. In the article a low-cost oriented system of gift cards authorization was described, with a special attention to social and organizational aspects.

12.1 The Business Environment of Gift Cards

Building permanent relation with customer is becoming one of the basic elements of competitive advantage and building strong market position. Educating customer the way it generates his returns and additionally causing his satisfaction and engagement in relation with the company is almost a guarantee of a profit and survival. Research shows that one loyal customer can create even 100 times more profits than potentially new customer. This is the result of the fact that the cost of acquiring and keeping customer in the longer period of time is very high. But the companies often forget about this fact. They invest huge amounts of money in signing new contracts, forgetting about the clients who already exist.

As an example we can give big corporation of mobile telephony or chosen petrol station networks. After time, during the boom of creating long lasting relations there

Joanna Zukowska
Karol Adamiecki University of Economics in Katowice,
Department of Enterprise Management
e-mail: joanna.zukowska@ae.katowice.pl

Zdzislaw Sroczynski
Institute of Mathematics,
Silesian University of Technology
e-mail: zdzislaw.sroczynski@polsl.pl

E. Tkacz and A. Kapczynski (Eds.): Internet – Technical Development and Appl., AISC 64, pp. 101–108.

is a reflection and a bow towards the permanent customers. The profit is usually on both sides, organization knows what customer expects and can easily adjust to his expectations, all suggestions are only constructive criticism helping innovation and development, the other side knows what kind of service can be expected. It is aware of the support which should be received. It is worth investing in permanent clients due to the fact that only satisfied customer can encourage three others on the other hand dissatisfied customer can give often negative opinion to even up to ten other potential buyers. Loyal customers are very good in the use of more and more common whisper marketing. Positive opinion on discussion forum is almost priceless for the company. Customer for years it is a prove of quality, brand, satisfaction, right choice. It is invaluable asset in the profit generation process.

We can name several advantages of the loyal customer, but before that several conditions must be met. First of all flawless service, satisfying of concrete needs, suitable advice and satisfactory post-sale service. The observation shows that post-sale service, customer care after closing sales is almost a guarantee of the next transaction. Of course in case of B2B market the situation is a bit different. Competitiveness is not as strong as in case of individual customer markets. In this case the most important factors are rational elements, measurable, references, company's reputation, its economic stability. Despite of all in both cases of actions connected with keeping the customer are appreciated. Except for the rebate programs, integration meetings, small company gifts longterm loyalty programs are becoming more important. Especially that the research show that 50% of customers who move to the competition are satisfied customers [6]. It is a proof that the important element is a search for permanent bounds with buyers.

12.1.1 Good Vouchers and Gift Cards Survey

One of the tools of customer relation marketing are goods vouchers. They are offered by more and more companies all over the world [7] and in Poland for example by Sephora, Duka, Douglas, Orsay, Neosport, Euroflorist, Silesia City Center, Real, different Spa & Wellness, Lynx Optique.

Moreover it is worth to pay attention to specialized companies, which offer gift cards/vouchers like for example Voucher Express Polska or Sodexho Polska. Voucher Express is market leader in the scope of gift cards. They are perfect motivation tool. Voucher Express services are used by HMV, Habitat, GAP, John Lewis, Marks & Spencer, Selfridges, House of Fraser, Harvey Nichols. Based on the opinion of people running the company the most popular and the most complex solution is

"SayShopping - multioptional gift voucher accepted in 12 000 the most popular and the most recognizable British retail shops among them in Comet, HMV, Debenhams and Boots. This universal voucher can also be used for service purchase, for entertainment and in alternative retail points such as Threshers, Theatre Tokens, JJB Sports, and even Cadbury Gifts Direct [9]."

Also home companies offer multicriteria solutions. For example Euroflorist offers a voucher to over fifty sport clubs in the country from the standard clubs to the extreme sport centers [10]. Sodexo takes second place in the scope of gift cards. It embraces several thousand companies in Poland. It is a no pay contribution often used as a Christmas, holiday [8] or occasional gift being the element of the bonus-motivation system and alternative to companies gadgets. What is interesting during the Christmas time the companies give very attractive discounts for purchasing vouchers. It is a part of loyalty package.

12.1.2 Formal Regulations of Gift Cards

We must take under consideration also law aspects during gift cards distribution:

- the value of gift vouchers financed from the Company's Social Contribution Fund does not cause the increase of the Social Security contribution,
- the value of gift cards given to employees is treated by the company as a cost of income acquisition,
- transfer of the gift vouchers to the employees does not cause the tax point according to the Value Added Tax regulations [11],

Moreover according to the VAT bill voucher is money substitute. Voucher transfer is neither a service nor a good transfer and does not have to be VAT taxed [12].

Gift cards usually bear values of 10, 15, 20, 50, 100, 200 zlotys. This is determined by technical conditions. Gift voucher has transaction code which must be given to the receiver. It also common to have a gift card as printed document [13]. The regulations of gift vouchers vary widely from one issuer to another. As mentioned before they are legal tenders but their acknowledgement is not a duty but a will of the seller. They are not printed by Central Bank so they are contractual unit of the exchange transactions. It bears hologram, series, register number and validity date. The vouchers are usually valid for no longer than 6 months. They cannot be exchanged for cash or changed and you cannot get the change from them in traditional cash. If damaged, with illegible series number or without hologram they are not accepted. They can be accepted as a payment mean in case when the value of purchased goods equals the nominal value of the gift card, when the value of purchased goods exceeds the nominal value of the voucher, the value of the goods is lower than nominal value of the voucher.

12.2 Gift Cards Authorization System Requirements

Some key features were determined by the authors during interviews with potential users of the system [4]. The most important was to develop the system as cost-effective, not demanding extra hardware nor new telecommunication services. Therefore, the following factors should be taken into consideration:

- ease of implementation at home versions of Windows operating system, common in small retails, where computer is used as a terminal for a fiscal printer,
- possibility of assignment a different value to each gift card,
- possibility of generating short series of unique gift card numbers,
- very low cost of remote communication with system, without Internet connection, prefferably with the use of GSM short messages (SMS),
- working in distributed network of retails, with joint card numbers database,
- no need to buy extra computer (server),
- implementation with use of popular cellular phones (GSM terminals), connected with USB cable,
- user-friendly interface for non-experienced users of the software, mostly shop-assistants,
- confidentiality of stored information regarding card numbers and values, authorization dates etc.,
- ensure the security of each transaction with just unique card identifier, independent from normal, but rather expensive, safeguards as holograms, embossed or convex parts of plastic cards, or watermarks and metal threads on paper cards,
- possibility to authorize gift card's value through the Internet service in case of eventual implementation of the Internet store.

The main target for described system are trade networks consisted of a few stores, issuing a few dozen of vouchers a year. The cost of implementation of out of the box solutions is unacceptable in this segment of the market. Almost every loyalty system for small business available widely in Poland requires relatively large amount of involved customers, formalities and formal contracts with providers.

To meet all requirements stated above, a system based on central database and authorization with the use of GSM messages was developed. The main advantages of this system are low implementation cost, operating in off-line software environment and possibility to issue card number in electronic form. Electronic gift card is just a numerical code, which can be sent by SMS, e-mail or even spelled into a voice mail or over the phone. The process of authorization of the card does not require any human interaction, therefore there is no need of any extra employees. This is very important for small and medium-sized businesses, where employment costs are significant. On the other hand, spreading one's business directs the interest towards Internet e-commerce. So, the described system is ready to accept cards' authorizations from e-shops.

12.3 System Architecture

The main components of the system are:

1. MS Windows GUI application residing in tray area, accessible for the operator issuing cards' tokens and checking validity of tokens during the payment,

2. SQL database system providing user list and generated gift cards' numbers/ tokens. Database primary key integrity assures unique value of each token — the chosen database system is open-sourced Firebird [5],
3. GSM terminal installed as a modem in the Windows operating system, connected through USB cable,
4. Communication service from GSM provider with SIM-card from voice contract or even pre-paid.

12.3.1 Gift Card Manager

The administrative Windows application was developed using Codegear Delphi and build-in database access components – dbExpress [3]. Open-source class DBMS ensures reliability, independence from hardware platform and scalability till potential full exchange with another database engine. DbExpress gives similar level of independence, based on SQL driver architecture with connectors to almost every kind of popular database server.

Tasks performed by dedicated win32 platform application (see Figure 12.1):

- generate unique tokens for new gift cards and store them in database,
- manual validation of a given card number/token,
- browsing across issued gift cards, restricted for administrators with complete information about card number (i.e. unique token), assigned value (*Polish zloty*), date and time of issue, date and time of the authorization, phone number of the mobile which asked for an authorization, user name who has created the card.

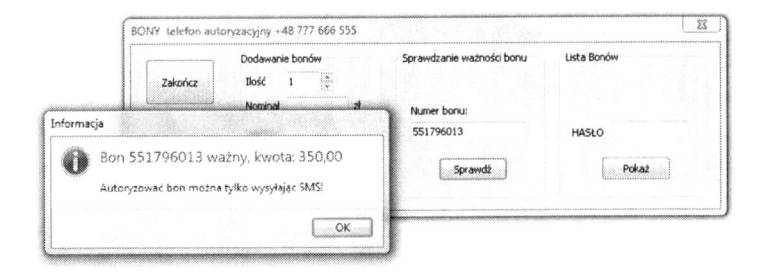

Fig. 12.1 Example of GUI of the Gift Card Manager application during checkout of given token number

In order to avoid frauds, cards' numbers are shown only to issuer (in the message box just after creation) and to the administrator. This protects confidential information about tokens, their numbers and values, which cannot be noted by ordinary shop assistant by chance. The authorization is possible only through SMS message sent from mobile terminal. Such procedure ensures binding between telephone number (rather personal data) and authorization process.

The application checks SMS memory of the connected phone periodically, decodes any new message, sends appropriate SQL queries to the database server and sends back an answer to the asking mobile. Clear information about token validity with exact time stamp and value confirmation are included in the answer.

12.3.2 Authentication Procedure

The protocol used in described system is very simple and consists of two SMS commands, which can be sent from asking mobile to authorization mobile number (see Figure 12.2):

1. CARDCHK <card_number>
this command[1] checks the validity of card number/token and sends back the value assigned to the card,
2. CARDAUTH <cardnumber>
this command[2] signs given card number as authorized (used), saves time stamp and number of the mobile which has sent the request into the database and finally sends SMS message with confirmation.

Thanks to these commands, the shop assistant can make sure that: gift card is valid, card was marked as used and it will not be possible to use it again, respectively.

Total cost of ownership for GSM-based authorization system is very low. The user (trade network) needs to purchase one more mobile terminal and not expensive license for the software. Running costs are equal to cost of sending a few SMS messages (amount depends on how many checks were made before final transaction).

The implementation of this system involves only one shop or a computer in headquarters (which can run only in opening hours), and so it is perfect for small and medium-sized businesses. Staff in the other shops need to know just authentifying phone number and the syntax of the SMS commands. In current fast changing employment market it is another important factor, because the training of the new workers is easy and fast.

12.3.3 SMS Protocol

The GSM terminal used by the system should be compliant with GSM Technical Specification [1][2], in particular it should perform SMS AT commands in text and PDU (*Protocol Description Unit*) mode. GSM phone operating as modem is controllable with special subset of AT commands. The PDU mode allows to send binary information in 7 bit or 8 bit format. The SMS message described by the ETSI organization, can be up to 160 characters long. The PDU command string (encoded

---

[1] Originally there is a Polish abbreviation BONSPR instead.
[2] Originally there is a Polish abbreviation BONAUT instead.

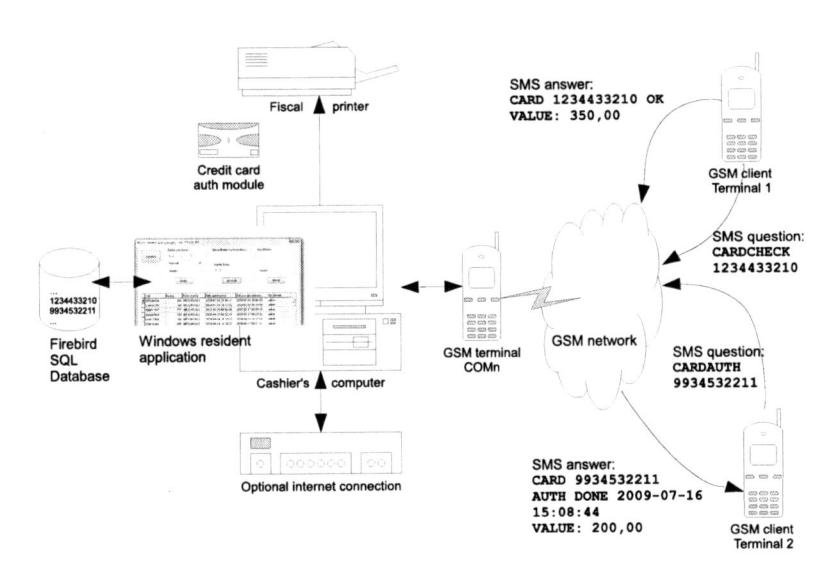

Fig. 12.2 The diagram of information flow in the system. Two SMS commands are shown with respective answers from the authorizing server

hexadecimal) contains not only the message, but additionally a lot of meta-information about the sender, his SMS service center, the time stamp etc. The incoming SMS string consists of initial byte indicating the length of the SMSC information, the SMSC data, and the SMS_DELIVER part. The main part includes sender number, protocol identifier, data coding scheme, time stamp, length of message, message body. The outgoing SMS control string is analogous, but of course there is destination phone number and validity-period flag.

During system tests some distinctions in implementation of the PDU mode in different GSM terminals were noticed, even when consider the same brand. These problems were solved by patching software communication library for particular GSM terminal model. On the other hand, some phones support SMS, but they use a proprietary protocol, which is not supported by used software libraries. Therefore, it seems important to extend gift card manager software with driver-based architecture to easy change the GSM phone that is used by the system. This way any terminal which meets the basic requirements would be suitable and will not require hard-coded changes in the software.

12.4 Conclusions

The system developed by the authors and described here allows to extend payment methods in small and medium-sized retail trade businesses. Gift cards made in a flexible way, secure, with varying values and easy authorization, help to develop and maintain good relations with clients. Our system is efficient enough to serve a few retail points and even thousands of vouchers. The total cost of ownership is affordable in comparison with available at Polish market, out of the box solutions.

References

1. ETSI GSM 07.05: Digital cellular telecommunications system (Phase 2); Use of DTE-DCE interface for Short Message Service (SMS) and Cell Broadcast Service (CBS)
2. ETSI GSM 07.07: Digital cellular telecommunications system (Phase 2); AT command set for GSM Mobile Equipment (ME)
3. Jensen, C.: Data Access Options with Borland RAD Tools. In: Borland Conference (2003)
4. Sroczynski, Z.: Gift Card Menager (BONY) internal technical documentation (2009)
5. Firebird SQL Database Documentation,
   `http://www.firebirdsql.org/index.php?op=doc`
6. Sikora, D.: Budowanie relacji z klientem (2007),
   `http://www.przepisnabiznes.pl/obsluga-klienta/`
   `relacje-z-klientem.html`
7. Simpson, S.: Are Gift Cards the Best Gift?,
   `http://ezinearticles.com/?Are-Gift-Cards-the-Best-Gift?&id=`
   `20869`
8. Faircloth, K.: Gift Cards Still Tops (2008),
   `http://www.inc.com/news/articles/2008/10/`
   `holiday-spending-gift-cards.html`
9. Voucher Express PL Web Site, `http://www.voucherexpress.pl`
10. EuroFlorist Web Site,
    `http://www.twojekwiaty.pl/do/page/bony_adrenalina`
11. Sodexho Web Site,
    `http://www.sodexho.pl/plpl/uslugi/MotywPracow/KupPodar/`
    `KupPodar.asp`
12. Bony podarunkowe a VAT,
    `http://centrumpr.pl/artykul/bony-podarunkowe-a-vat,4991.html`
13. `http://www.neosport.pl/Bony-podarunkowe-cterms-pol-20.html`

Chapter 13
Remote Access to a Building Management System

Krzysztof Dobosz

Abstract. This paper presents a universal approach to the problem of a remote access to the building management system. The idea bases on the configuration of the BMS stored in the XML format. The configuration consists of defined properties of buildings, devices and states. The building management protocol for remote access to the BMS is also described. Proper commands for controlling devices, monitoring are enumerated. Shortly the solution of the problem of transaction is presented.

13.1 Introduction

BMS (Building Management Systems) is a system for the integrated management of all the technological functions of a building [1, 2]. These allow to monitor and to manage all key building parameters – from power supply to security access [3]. Features of the BMS can be as follows: monitoring and management of power, cooling and humidity, customer infrastructure interface to monitor individual room systems, monitoring power for consumption statistics and billing, leak detection from cooling systems, generation of system performance and facility data, audio and visual alarms, habitants access control, etc.

Complex BMS can simulate habitants' presence as well. An infrastructure of the BMS consists of different devices, which can be monitored and managed, i.e.: audio-visual devices, illumination devices, electromagnetic locks, doors, windows, gates, cameras, sensors and different detectors.

Generally devices can be divided to passive and active. Passive devices allow to get information about their current state. Active devices additionally can be managed by setting their states. Buildings, devices, and states have to be configured in

Krzysztof Dobosz
Faculty of Automatic Control, Electronics, and Computer Science,
Silesian University of Technology, Gliwice
e-mail: krzysztof.dobosz@polsl.pl

E. Tkacz and A. Kapczynski (Eds.): Internet – Technical Development and Appl., AISC 64, pp. 109–117.

the same way, in spite of their technical differences. But not only the infrastructure of the BMS is important, also the degree of remote managing that can offer. Devices can be controlled via cable or wireless connection as well as from local server console. Client software can be implemented as a standalone application, web component, application embedded in a remote control or mobile phone. Integrated remote management for building facilities can provide convenience, comfortable software, which allows managing information on the BMS, and directly retrieving necessary information at any time. Realizing different client software we need a common protocol for communication.

13.2 Configuration of the BMS

Before the connection, the remote client has no information about the BMS. After the first connection, the control gets the information about all devices. Configuration of the BMS should be stored with the universal data format using XML tags.

13.2.1 Buildings, Devices, States

The main tag in the BMS description file is `<bms>`. This tag consists of following sub-tags:

- `<description>` – description of the BMS system,
- `<ssl-support>` – values yes/no determine possibility of encoding the communication,
- `<password>` – administrator password; if empty - access to every building is separated,
- `<buildings>` – list of `<building>` tags,
- `<building>` – main tag of definition in a single building.

Every building is obligatory described with two tags:

- `<building-id>` – building unique identifier,
- `<building-devices>` – list of `<device>` tags.

Optionally it can include also other tags, i.e.:

- `<building-name>` – building short name,
- `<building-description>` – building long description,
- `<building-password>` – password protecting from the access to the building configuration.

Every monitored or managed device is described with the following tags:

- `<device>` – main tag of a single device definition,
- `<device-id>` – device unique identifier,

- `<device-type>` – type `passive/active`, a passive device has no states to set, can be only monitored,
- `<device-recorder>` – value `yes/no` carry the information on an additional binary data, `yes` – indicates a device running period of time as camera or sound recorder, which acquire binary data and store it in the BMS file system or database,
- `<device-states>` – tag grouping repeatedly `<state>` tags.

Above tags are required for correct configuration of the BMS. Additionally `<device>` tag can contain more information described by:

- `<device-name>` – short device name,
- `<device-description>` – long device description,
- `<device-icon>` – path to the file localized on the BMS server, containing graphical representation of the device.

Every device state can be defined with two obligatorily tags:

- `<state>` – main represents single device state,
- `<state-id>` – unique state identifier.

As in case of the building and the device, also that can contain additional tags, i.e:

- `<state-icon>` – path to the graphical representation of the current state,
- `<state-name>` – short state name,
- `<state-description>` – long state description.

The solution using optional tags is very elastic, because the BMS administrator can define new attributes for special properties of buildings, devices and states. The format of a new tag for a building is `<building-attr>`, where `attr` represents attribute.

13.2.2 Automation

BMS configuration file also includes information about periodically changed device states, and relations between states of different devices. Indicated states in defined period are set automatically. This part of the configuration file begins with the tag `<actions>`. This tag consists of the sequence of `<action>` tags. Subtag `<action>` defines a single relation between two devices: signal sender and signal receiver. Sender generates the signal changing a state to the defined as a signal source. The signal is forcing to change a state of receiver to the defined as a signal destiny. The `<action>` tag includes two sub-tags: `<signal-source>` and `<signal-destiny>`. Proper XML tags for both are: `<building-id>`, `<device-id>` and `<state-id>`.

13.2.3 Events Dictionary

BMS events dictionary uses XML format. For remote client this file can be opened only in the read-only mode. Main `<event-dictionary>` consists of many `<event>` tags. Every one consists of following sequence, reporting about a new device state: `<date>` – date stored in the format: `yyyy.mm.dd`, `<time>` – time stored in the format: `hh:mm:ss`, `<building-id>`, `<device-id>`, `<state-id>`.

13.3 Building Management Protocol

The most popular BACnet protocol defines a number of services that are used to communicate between building devices [4, 5]. Some solutions extend that idea [6]. Other approaches concentrate on web access [7] or wireless communication [8]. Proposed BMP (Building Management Protocol) is the protocol located in application layer of OSI model. It is similar to the FTP protocol. BMP is dedicated to remote communication and it is independent of an Internet transport media. It consists of 20 commands. They are grouped as follows: base commands, information commands, controlling commands, and monitoring commands. Every command is a key word built with four letters. This way simplifies commands processing in the BMS.

Execution of every command is confirmed by the BMS. A confirmation message consists of a single line. The line starts with the message code and finishes with the message body. Messages codes are placed in two ranges:

- 100–199 – commands realized,
- 200–299 – commands rejected.

This solution simplifies messages processing on the client device side.

13.3.1 Initialization and Finishing

Base commands realize access to the infrastructure description stored in the `bms. xml` file. A session is initiated when a client opens a connection to a BMS server and the server responds with an opening message:

```
100 Welcome to the BMS. System date is: yyyy.mm.dd,
system time is: hh.mm.ss
```

The first step of using the BMS service is to send the invitation command:

```
HELO client-id
```

This command is used to identify the client to the BMS server. Command parameter should be IP number in the case of TCP connection, device address in the case of Bluetooth connection, etc. The BMS confirms using the message:

```
101 Hello <client-id>
```

Next the client device has to be authenticated. It should send following commands:

```
USER user
PASS password
```

Correct pair user-password is confirmed with the message:

```
102 User is authenticated
```

Sending incorrect pair of data finishes with the message:

```
201 Wrong user or password. Access to the
BMS is permitted
```

Any other commands is not executed, and the BMS every time sends:

```
202 Command canceled. User is not authenticated
```

After the correct authentication, the client can send other commands. The QUIT command specifies that the receiver must close the transmission channel. Every BMS server should also implement the HELP keyword as well. This command causes the server to send list of helpful information to the client.

13.3.2 Information Commands

Information commands realize access to the infrastructure description stored in the bms.xml file.

- LSTB, LSTD, LSTS – return information about elements defined in the infrastructure: buildings, devices and states; the list of devices concerns the current building; the list of states concerns the current device,
- GETB attr – returns attribute (id or any optional) of the current building,
- GETD attr – returns attribute (id, type, recorder or any optional) of the current device,
- GETS attr – returns attribute (id or any optional) of the current state.

Commands: GETB, GETD, GETS without a parameter, return whole list of defined attributes. Every information command is confirmed with one of the messages:

```
111 Requested information is sent
211 No such attribute
```

13.3.3 Controlling Devices

Information commands realize access to the infrastructure description stored in the bms.xml file.

- SETB id – change the current building to the another one represented by id,
- SETD id – change the current device to the another one installed in the current building, and represented by id,
- SETS id – change the current device state to another one, defined for the current device, and represented by id.

The above commands can be omitted, and then the first defined element (building, device or state) in the bms.xml file, is treated as the current. Direct change from the current state to the new one can be impossible, and then are used transitional states. That will be reported in the events' dictionary. Every controlling command is confirmed with the one of messages:

```
121 Current building is <building-name>
122 Current device is <device-name>
123 Current device is set to the state: <state name>
221 No building has such ID
222 No device in the current building has such ID
223 No device state has such ID
224 The state is not changed.
    Device <device-id> is passive
```

13.3.4 Monitoring

Monitoring commands realize access to the information about events stored in the event dictionary file. The client can use one of three commands:

- HIST date time – changes the indicator of the event history to the date and time. Date format is yyyy.mm.dd. Time format is: hh:mm:ss. Parameters date and time can be omitted, then default values are: current day, current BMS time,
- GETE number – returns indicated number of last events, counting from the adjusted indicator backwards; number can be omitted – then information about the only one event is returned,
- GETB – returns a sequence of binary data recorded by the current device and stored in the BMS file system or database, The sequence finishes with the server confirmation message (code 132).

The format of the line in response for GETE command is:

```
yyyy.mm.dd hh:mm:ss: Building <building-id>,
device <device-id>, new state is <state-id>
```

After sending information about requested number of events, the command is confirmed with one of messages:

```
131 Events are reported
132 Binary data is sent
231 Invalid date/time format
232 Device is not a recorder
```

13.3.5 Transactions

The idea of BMS allows to the simultaneous access of many clients to the system. It isn't significant, whether many customers want to manage devices at the same time. Every command by default determines separated transaction, but that can be switched off. Command sequence can be treated as one transaction, when starts with the TRNS command and finishes with COMM command. TRNS command is confirmed with the message

```
141 Transaction is opened
```

or

```
241 Transaction is already open
```

COMM finishes the sequence of commands grouped in the same transaction. In case of realization of all commands by the BMS, this command is confirmed with message:

```
142 Transaction is committed
```

In case when at least one command cannot be realized, the BMS rollbacks all commands and sends the message:

```
242 Transaction is rejected
```

Using the transaction command shows the example below:

```
HELO <client-id>
USER <user>
PASS <password>
TRNS
SETB office
SETD mainDoor
SETS close
SETD mainLamp
SETS switchOff
SETD sensor
SETS on
COMM
QUIT
```

13.4 Remote Access

We assume the both synchronous and asynchronous remote access to the BMS. The BMS should be independent of the access type and independent of used protection for sent data. BMP commands can be transmitted from a client to the BMS by different providers. Then, on the BMS server communication is realized through the plug-ins group. Every plug-in should implement common BMS Server API and uses different transmission medium, i.e. Internet, radio channel, Bluetooth, SMS, etc. The role of the plug-in is to transfer a BMP command to the BMS engine. In such way communication module is independent of a kind of remote access.

The synchronous access bases on such low-level connection protocol as TCP or Bluetooth. Both have important for the BMS features:

- ordered data transfer – the destination host rearranges according to sequence number,
- retransmission of lost packets – any cumulative stream not acknowledged will be retransmitted,
- discarding duplicate packets,
- error free data transfer.

Using TCP, implementation of SSL for data encryption should be done in a proper plug-in. Bluetooth implements authentication and key derivation with custom algorithms. In Bluetooth, key generation is generally based on a Bluetooth PIN, which must be entered into both devices. During pairing, an initialization key is generated, using a proper algorithm. This key is used for subsequent encryption of data sent via the air.

SMS (Short Message Service) unit can be treated as a batch script message. Implementation should require server response in a form of the SMS as well. Confirmations should be sent back in the SMS form. Asynchronous communication is dedicated to the transaction mode.

Developers should pay particular attention to the asynchronous communication, because the SMS standard does not provide encryption. In case of access protected by a password, using the BMS system won't be safe.

13.5 Conclusions

In the paper different aspects of remote access to the BMS server are discussed. At the beginning BMS configuration file and event dictionary is shortly described. Then Building Management Protocol is presented and their commands are enumerated. The BMS independence of applied low-level transfer protocol is also underlined.

Proposed format of the BMS configuration file, event dictionary file and the BMP protocol can be also used for monitoring and managing with almost everything, for example with computer network and their software. Personal computer can play the role of a building, an application – a role of a device, an application command – a state of a device.

References

1. Pennycook, K.: Standard specification for BMS. The Chameleon Press Ltd. (2001)
2. Hai-young, J., Sang-chul, Y., Duck-gu, J.: Building Management System and Method. U.S. Patent Application 20080195687 (2008)
3. Cser, J., Beheshti, R., van der Veer, P.: The development of an Integrated Building Management System. In: Portland International Conference on Management and Technology (1997)
4. BACnet - A Data Communication Protocol for Building Automation and Control Networks. Official Website of ASHRAE SSPC 135, http://www.bacnet.org/
5. Bushby, S.T.: BACnet: A Standard Communication Infrastructure for Intelligent Buildings. Automation in Construction 6, 529–540 (1997)
6. Tazaki, S.: A Standard Protocol for an Autonomous, Decentralized Building Automation System. In: Proceedings of the The Fourth International Symposium on Autonomous Decentralized Systems. IEEE Computer Society, Los Alamitos (1999)
7. Shengwei, W., Junlong, X.: Integrating Building Management System and facilities management on the Internet. Automation in Construction 11(6), 707–715 (2002)
8. Gylling, M.: Remote wireless control of building management systems automation. Master Thesis, Blekinge Institute of Technology, Ronneby (2004)

Chapter 14
Applying Rough Set Theory to the System of Type 1 Diabetes Prediction

Rafał Deja

Abstract. In this paper we present the results of analysis of the genetic data of children with diabetes mellitus type 1 (DMT1) and their up to now healthy siblings. Base on the data the decision support system has been developed to classify and later on to predict the illness among the children with genetic susceptibility to DMT1. The system can recommend to include a person to pre-diabetes therapy. The data mining and classification algorithms are based on the rough set theory.

14.1 Introduction

It is well known that type 1 diabetes is a disease with the genetic background. Many authors [3] has shown that the risk of developing the disease is much more higher among the first degree relatives than in the population (0.4%) and is evaluated accordingly: for siblings 6%, for sick mothers offspring $3 - 4\%$ and for sick fathers offspring $6 - 9\%$. At the same time the above-quoted results mean that not all children with genetic susceptibility to DMT1 will ever develop the disease. Epidemiological data provide evidence to link some environmental triggers with DMT1 (e.g. viruses). A lot of genes are responsible for developing the disease however the most important genetic susceptibility to DMT1 is conferred by genes constituting the HLA complex, located on the short arm of chromosome 6. Additionally the genetic predisposition of $Th1$ and $Th2$ lymphocytes cytokines' production can introduce the onset of the illness, thus these genetic results has been consider either.

Rafał Deja
Department of Computer Science,
Academy of Business in Dabrowa Gornicza,
Cieplaka 1c, 41-300 Dabrowa Gornicza
e-mail: rdeja@wsb.edu.pl

E. Tkacz and A. Kapczynski (Eds.): Internet – Technical Development and Appl., AISC 64, pp. 119–127.
springerlink.com

The aim of the study was to develop the decision support system. The information system allows the classification of children with genetic susceptibility to DMT1 to those with higher and lower risk of falling ill. The children from the first category can be carefully treated by doctors including apoptosis of beta cells prevention. The arrangement of genes responsible for disease development is to be deducted from the data.

The data collected in the research like other medical data are complex, sometimes uncertain and incomplete. Furthermore the relations among the attributes are hard to be described. Thus we decided to use algorithms and methods based on the rough set theory. In many papers [6, 11, 12, 13, 14] the suitability of these methods in medical analysis has been proven.

14.2 Medical Data

Clinical material analyzed in the paper consists of:
- 44 children with diabetes mellitus type 1 (the average age is 9.82)
- Their healthy siblings – also 44 cases (the average age is 10.64)
- The healthy control group – 36 persons without any symptoms of illness and genetic predisposition in medical history.

All of the patients were examined by taking the blood sample and making genetic tests.

14.3 The Aim of the Study

The main goals of the study are as follow:

1. Building decision support system for DMT1 prediction among the genetically loaded children.
2. The analysis of the genes arrangement that can yield to sickness. The genetic differences between the sick children and they healthy siblings and the healthy control group.
3. Induction of the classifier of genetic susceptibility to DMT1 based on the rough set theory.
4. Appling the classifier to the healthy siblings and indicating the children with higher risk of developing the disease.

We are expecting that the genes arrangement of the healthy siblings must be strongly close to that of the children with diabetes. However there must be a difference that cause some of the relatives to stay healthy, and the others to fall ill. The final classifier that builds the decision support system defines the genes arrangement that increase/decrease the risk of onset of DMT1.

14.4 Rough Set Preliminaries

Rough set theory was developed by Zdzislaw Pawlak [9] in the early 1980's. It is an important tool to deal with uncertain or vague knowledge and allows clear classificatory analysis of data tables.

An information system is a pair $S = (U, A)$, where U - is nonempty, finite set called the universe; elements of U are called objects, A - is nonempty, finite set of attributes. Every attribute $a \in A$ is a map, $a : U \rightarrow V_a$, where the set V_a is the value set of a; elements of V_a.

A decision system is any information system of the form $I = (U, A \cup d)$, where $d \notin A$ is called the decision attribute. The elements of A are called conditional attributes.

With any not empty set $B \subseteq A$, we define the equivalence relation called B-indiscernibility:

$$IND(B) = (x, y) \in U \times U : \forall a \in B\, (a(x) = a(y)).$$

Objects x, y that belong to the relation $IND(B)$ are indiscernibile from each other by attributes from B. The minima subset $B \subseteq A$ such that $IND(A) = IND(B)$ is called the reduct of set A. The set of all reducts from A is denoted by $RED(A)$.

The relation $IND(B)$ introduces the division of all objects from U into equivalence classes

$$[x]_B = y : x IND(B) y \quad \text{for each} \quad x \in U.$$

The sets $\underline{B}X = \{x \in U : [x]_B \subseteq X\}$, $\overline{B}X = \{x \in U : [x]_B \cap X \neq \varnothing\}$ are called the B-lower and B-upper approximations of X, respectively. The set $\underline{B}X$ is also called positive region of X. The accuracy of approximation can be measured with coefficient $\alpha_B(X) = \frac{|\underline{B}X|}{|\overline{B}X|}$, where $|X|$ denotes the cardinality of X. Obviously $0 \leq \alpha_B(X) \leq 1$. If $\alpha_B(X) = 1$, X is crisp with respect to B, and otherwise, if $\alpha_B(X) < 1$, X is rough with respect to B.

14.5 Information System

The analysis and experiments were conducted using RSES (Rough Set Exploration System) system [2]. The system (current version is 2.2) has been created by so called Group of Logic at the Warsaw University http://logic.mimuw.edu.pl/ under the supervision of Andrzej Skowron. The most important algorithms and methods of rough set theory has been implemented there. More precisely we make use of the algorithms for reducts calculation, for rules generation especially the algorithm LEM2 in this study.

14.6 Data Analysis

In the study the attributes presented in Table 14.1 has been considered.

Table 14.1 The attributes used in the study

Attribute (A)	Card. of value set of A	Medical meaning
DRB1	24	Allele of class II subregion of DRB of the HLA complex on the first chromosome
DRB2	24	Allele of class II subregion of DRB of the HLA complex on the second chromosome
DQB1	12	Allele of class II subregion DQB of the HLA complex on the first chromosome
DQB2	13	Allele of class II subregion DQB of the HLA complex on the second chromosome
TNF	2	TNF-alpha gene -308 A/G polymorphism on the first chromosome
TNF2	2	TNF-alpha gene -308 A/G polymorphism on the second chromosome
IL10	4	Interleukin-10 gene polymorphism on the first chromosome
IL102	3	Interleukin-10 gene polymorphism on the second chromosome
IL6	2	Interleukin-6 gene polymorphism on the first chromosome
IL62	2	Interleukin-6 gene polymorphism on the second chromosome
IFN	2	Interferon gene polymorphism on the first chromosome
IFN2	2	Interferon gene polymorphism on the second chromosome

At the very beginning, using the results of genetic tests, three decision tables has been created. The first decision table describes the children with diabetes and healthy control group, the second children with diabetes and they healthy siblings, and the third healthy siblings and healthy control group and are denoted by $DT1$, $DT2$ and $DT3$ respectively.

The reducts and rules sets from the decision tables were calculated. Reducts are calculated directly using the discernibility matrix. The exemplar decision table $DT2$ of children with diabetes and they healthy siblings is presented in Table 14.2. The decision $D = 1$ denotes ill children and $D = 0$ denotes healthy sibling.

Reducts calculated from decision table children with diabetes vs. healthy sibling are as follows:

RED(A) = {{DRB1, DQB1, TNF, IL6, IFN, DR2, IL102, IFN2}, {DRB1, TNF, IL6, IFN, DRB2, TNF2, IL102, IFN2}}

In Table 14.3 we collected the results for all decision tables.

Table 14.2 Decision table TDM1 children vs. healthy sibling (fragment)

DRB1	DQB1	TNF	IL10	IL6	IFN	DRB2	DQB2	INF	IL102	IL62	IFN2	D
301	201	A	ATA	G	T	1601	501	A	ATA	G	T	1
701	201	G	GCC	G	T	301	201	G	ACC	C	A	1
1601	502	G	ACC	G	T	401	302	A	ACC	C	T	1
401	302	G	GCC	G	T	301	201	A	ACC	C	A	1
701	202	G	GCC	C	T	404	302	G	ATA	C	T	1
1502	602	G	GCC	C	A	1601	501	G	GCC	C	A	0
701	201	G	GCC	G	T	701	201	G	GCC	C	A	0

Table 14.3 The results comparison for all 3 decision tables

Decision table	Reducts no	$\propto$	Dispensable attributes	Rules no
Children with diabetes vs. healthy control group (DT1)	62	1	None	4461
Children with diabetes vs. healthy siblings (DT2)	2	0.88	IL10, DQB2	162
Healthy siblings vs. healthy control group (DT3)	44	1	TNF	3317

One can noticed directed by the number of reducts and the accuracy of approximation coefficient that the sick children and they healthy siblings are similar when the genes arrangement is considered. In contrast the high number of reducts obtained from the decision tables $DT2$ and $DT3$ testify the high amount of equivalence classes. Taken into account the genes arrangement the persons from these groups are completely different.

14.6.1 Classification Problem

In typical scenario of solving the classification problem [1] the input data set is divided into two separated parts. Using one of the part, called the training set, the classifier is taught. The classifier is applied and verified using the remaining part, called the test set. The result of the process is the classification on the test set and the structure of the classifier that can be used as a decision system. This method is often called *train-and-test*.

Using the data we have gathered the classification process can be performed differently. As a training set for classifier to be deducted we use the decision table $DT1$ ($DMT1$ children vs. healthy control group). In the next step the classifier will be applied to healthy sibling set (our test set) the classification of which is desired. In the paper the rules classifier has been used.

The straight forward rules generation from entire reducts set can be inconvenient because of the possible huge number of rules (very detailed classification) – see Table 14.3. Moreover the rules power is low – e.g. for the rules generated from $DT1$

the maximal rule power is 2. Thus in the research described in the paper we decided to use *LEM2* algorithm for rules generation. The *LEM2* algorithm originally proposed by Grzymala-Busse [4, 5] consists on calculation of the local cover for each approximation of the decision table concept. It then convert them into decision rules. The algorithm has been successfully used in many medical problems [6, 11, 14].

Using the algorithm *LEM2* from the decision table children with diabetes vs. healthy control group the 38 rules has been deducted. These rules as classifier has been applied into the data table with healthy siblings. The classification outcomes are as follows: 10 children has been classified as healthy, 19 as sick, the remaining 19 could not be certainly classified as either sick or healthy. The classification results are partially provided in Table 14.4. The decision attribute *D* meaning is healthy or sick with value 0 or 1 respectively.

Table 14.4 Classification results (fragment)

DRB1	DQB1	INF	IL10	IL6	IFN	DRB2	DQB2	INF2	IL102	IL62	IFN	D
1502	602	G	GCC	C	A	1601	501	G	GCC	C	A	0
701	201	G	GCC	G	T	701	201	G	GCC	C	A	1
301	201	G	ACC	G	A	402	302	A	ACC	C	A	MISSING
1601	502	G	ACC	G	T	401	302	A	ACC	C	T	1
401	302	G	GCC	G	T	301	201	A	ACC	C	A	1
301	201	G	GCC	G	T	404	302	A	ACC	C	T	0
101	501	G	GCC	C	T	419	304	G	ATA	C	A	MISSING
404	302	G	GCC	G	A	1103	301	G	ATA	C	A	1
401	302	G	GCC	G	T	1601	501	G	ACC	C	T	1
701	202	G	GCC	G	T	402	302	G	ACC	C	T	1
101	501	G	ACC	G	T	401	302	G	ATA	G	A	0
1306	602	A	GCC	C	T	401	302	A	ATA	C	A	1
1601	502	G	GCC	G	A	101	501	A	ACC	G	A	MISSING

The Table 14.4 can be treated as decision table. Thus the rules describing potentially healthy and potentially sick children can be deducted. The *LEM2* algorithm has been used again, and the results are as follows (in the square brackets the rules power is given):

Potentially healthy children:
$(TNF = G)\&(IFN2 = A)\&(IL102 = ATA)\&(IL6 = G)\&(IL62 = G)$
$\Rightarrow (D = 0)$ [4]
$(TNF = G)\&(IL10 = GCC)\&(IL62 = C)\&(DRB1 = 301)\&(DQB1 = 201)$
$\&(IFN = T)\&(IFN2 = T) \Rightarrow (D = 0)$[3]
$(TNF = G)\&(IL10 = GCC)\&(IFN = A)\&(TNF2 = A)\&(IL102 = ATA)$
$\Rightarrow (D = 0)$ [2]
$(TNF = G)\&(IL10 = GCC)\&(IFN = A)\&(IL62 = C)\&(IFN2 = A)$
$\&(IL6 = C)\&(DRB1 = 1502) \Rightarrow (D = 0)$[1]
$(TNF = G)\&(IFN2 = A)\&(IL102 = ATA)\&(IL6 = G)\&(IL62 = G)$
$\Rightarrow (D = 0)$ [4]

$$(TNF = G)\&(IL10 = GCC)\&(IL62 = C)\&(DRB1 = 301)$$
$$\&(DQB1 = 201)\&(IFN = T)\&(IFN2 = T) \Rightarrow (D = 0) \text{ [3]}$$
$$(TNF = G)\&(IL10 = GCC)\&(IFN = A)\&(TNF2 = A)\&(IL102 = ATA)$$
$$\Rightarrow (D = 0) \text{ [2]}$$
$$(TNF = G)\&(IL10 = GCC)\&(IFN = A)\&(IL62 = C)\&(IFN2 = A)$$
$$\&(IL6 = C)\&(DRB1 = 1502) \Rightarrow (D = 0) \text{ [1]}$$

Potentially sick children:
$$(IL62 = C)\&(TNF = G)\&(IL6 = G)\&(IL10 = GCC)\&(TNF2 = G)$$
$$\Rightarrow (D = 1) \text{ [9]}$$
$$(IL62 = C)\&(IFN = T)\&(TNF2 = A)\&(IFN2 = A)\&(TNF = G)$$
$$\&(IL10 = ACC) \Rightarrow (D = 1) \text{ [4]}$$
$$(IL62 = C)\&(IFN = T)\&(DRB2 = 401)\&(DQ2 = 302)\&(TNF2 = A)$$
$$\&(TNF = A)\&(IL10 = GCC) \Rightarrow (D = 1) \text{ [2]}$$
$$(TNF = G)\&(IL62 = C)\&(IFN = T)\&(DRB1 = 1601) \Rightarrow (D = 1) \text{ [2]}$$
$$(DQB1 = 302)\&(TNF = G)\&(IL10 = GCC)\&(IL62 = C)\&(IFN2 = A)$$
$$\&(DRB1 = 401)\&(IL6 = G)\&(IFN = T)\&(DRB2 = 301)\&(DQ2 = 201)$$
$$\&(TNF2 = A)\&(IL102 = ACC) \Rightarrow (D = 1) \text{ [1]}$$
$$(DRB1 = 402) \Rightarrow (D = 1) \text{ [1]}$$

Finally we have deduced the rules classifier that describes the arrangement of genes that can indicate the genetic susceptibility to $DMT1$ (potentially sick children) and genetic resistance to developing the disease (potentially healthy children).

14.7 Conclusions

In the paper the building of the expert system has been presented. The system can support the medical doctors in predicting the disease among the healthy siblings of children with DMT1 or generally among the children with positive medical history. The attributes has been chosen using the expert knowledge.

The genes arrangement of the healthy siblings are strongly close to that of the children with diabetes according to both from HLA matrix and polymorphism of cytokine genes. However the differences has been deducted from data and the rules generated that describe the genes arrangement that can cause some of the relatives to stay healthy, and the others to develop the disease. One can interpret the rules as the genetic susceptibility to DMT1 or genetic resistance to fall ill. For example to the group of pre-diabetic children should be included these who are characterize by the alleles of HLA: $DRB1 = 402$ or $DRB1 = 1601$ and has the following polymorphism of cytokine genes $TNF = G$ and $IL6$ (at the second chromosome)$= C$ and $IFN = T$; etc. The children from that pre-diabetes group should be the subject of detailed medical supervision.

The way the classifier has been built allows the system to be inductively learned with new cases.

In the study the rough set theory algorithms and methods were used. Using this approach to data analysis has many important advantages like for example (see [7] for details):

- Identification of relationships that would not be found using statistical methods.
- Reduction of data to a minimal representation.
- Synthesis of classification or decision rules from data.
- Straightforward interpretation of synthesized models.

The algorithm *LEM2* has been intensively used throughout the research. The number or generated rules describing the classification has been restricted and became readable by physician. The power of rules increased. Nevertheless the classifier described a lot of cases uncertainly (15). The following investigations can be directed to improve the classifier. One of the possible solution is to reduce the value set of some of the attributes especially these from HLA matrix.

Acknowledgments

Data are obtained from the Pediatric, Endocrinology and Diabetology Department of Silesian Medical University thanks to courtesy of Grazyna Deja, MD, PhD.

References

1. Bazan, J., Nguyen, H.S., Nguyen, S.H., Synak, P., Wróblewski, J.: Rough set algorithms in classification problems. In: Polkowski, L., Lin, T.Y., Tsumoto, S. (eds.) Rough Set Methods and Applications: New Developments in Knowledge Discovery in Information Systems. Studies in Fuzziness and Soft Computing, vol. 56, pp. 49–88. Physica-Verlag, Heidelberg (2000)
2. Bazan, J., Szczuka, M., Wróblewski, J.: A new version of rough set exploration system. In: Alpigini, J.J., Peters, J.F., Skowron, A., Zhong, N. (eds.) RSCTC 2002. LNCS (LNAI), vol. 2475, pp. 397–404. Springer, Heidelberg (2002)
3. Deja, G., Jarosz-Chobot, P., Polańska, J., Siekiera, U., Małecka-Tendera, E.: Is the association between TNF-alpha-308 A allele and DMT1 independent of HLA-DRB1, DQB1 alleles? Mediators Inflamm. 2006(4), 19724 (2006)
4. Grzymala-Busse, J., Wang, A.: Modified algorithms lem1 and lem2 for rule induction from data with missing attribute values. In: Proc. of 5th Int. Workshop on Rough Sets and Soft Computing, pp. 69–72 (1997)
5. Grzymala-Busse, J.W.: Mlem2-discretization during rule induction. In: Proceedings of the International IIS, pp. 499–508 (2003)
6. Ilczuk, G., Wakulicz-Deja, A.: Rough sets approach to medical diagnosis system. In: Szczepaniak, P.S., Kacprzyk, J., Niewiadomski, A. (eds.) AWIC 2005. LNCS (LNAI), vol. 3528, pp. 204–210. Springer, Heidelberg (2005)
7. Komorowski, H.J., Pawlak, Z., Polkowski, L.T., Skowron, A.: Rough Sets: A Tutorial, pp. 3–98. Springer, Singapore (1999)
8. Midelfart, H., Komorowski, H.J., Norsett, K.G., Yadetie, F., Sandvik, A.K., Laegreid, A.: Learning rough set classifiers from gene expressions and clinical data. Fundamenta Informaticae 53, 155–183 (2002)

9. Pawlak, Z.: Rough Sets: Theoretical aspects of reasoning about data. Kluwer Academic Publishers, Boston (1991)
10. Skowron, A., Rauszer, G.: The discernibility matrices and functions in information systems. In: Słowinski, R. (ed.) Intelligent decision support. Handbook of Applications and Advances of the Rough Sets Theory, pp. 331–336. Kluwer Academic Publishers, Dordrecht (1992)
11. Slowinski, K., Stefanowsk, J., Siwinski, R.: Application of rule induction and rough sets to verification of magnetic resonance diagnosis. Fundam. Inform. 53, 345–363 (2002)
12. Tsumoto, S.: Extracting structure of medical diagnosis: Rough set approach. In: Wang, G., Liu, Q., Yao, Y., Skowron, A. (eds.) RSFDGrC 2003. LNCS (LNAI), vol. 2639, pp. 78–88. Springer, Heidelberg (2003)
13. Tsumoto, S.: Mining diagnostic rules from clinical databases using rough sets and medical diagnostic model. Information Sciences: An International Journal 162, 65–80 (2004)
14. Wakulicz-Deja, A., Paszek, P.: Applying rough set theory to multi stage medical diagnosing. Fundamenta Informaticae 54, 387–408 (2003)

Part II
Information Management Systems

Chapter 15
Outline of the Enterprise Resource Control Systems Architecture

Mirosław Zaborowski

Abstract. The tentative thesis of the Enterprise Resource Control (ERC) theory is the statement, that every planning and control system, taken from any enterprise, may be transformed with all its functions and data to an ERC system or to its part. In spite of such declared generality of the ERC theory the structure of the framework ERC system is relatively simple. All information describing organizational and functional structure of an ERC system, the structure of its data, as well as data values may be recorded in the relational database of this system. The key attributes of all tables belonging to any ERC system may be taken from the settled set of 20 structural attributes. The framework ERC system may have any number of organizational levels (typically four: primary organizational system, a plant, an organizational cell and a workstation). In each level, except of the lowest one, there are four functional layers of coordination, reengineering, allocation and execution planning. For formal description of ERC systems the Organizational Information-Transition Nets (OITN) are applied. The OITN structure is similar to the structure of Coloured Petri Nets (CPN). Pages from a hierarchical CPN correspond to organizational systems belonging to a multilevel OITN. OITN transitions are sites of data processing procedures. OITN places, incident to a given transition, are the sets of those tuples from the database tables, which are data inputs or data outputs of this transition.

15.1 Data Structure

The Enterprise Resource Control (ERC) systems are integrated planning and control systems with universal organizational, functional and informational structure. Data structure of these systems is like in relational databases [5]. Such data recording, in tables or in single records, is quite common in practice. It refers not only to

Mirosław Zaborowski
Academy of Business in Dabrowa Górnicza
e-mail: mzaborowski@wsb.edu.pl

E. Tkacz and A. Kapczynski (Eds.): Internet – Technical Development and Appl., AISC 64, pp. 131–139.
springerlink.com © Springer-Verlag Berlin Heidelberg 2009

relational databases, which are applied to most management systems, but also to databases of lower level control systems, as well as to memory of direct control devices. The tables of an ERC database, which are not subclasses of other tables, are called **kinds of administrative information**. Tuples in tables of information kinds are called **information elements**. Relationships between information kinds and their subclasses are modelled by means of E-R diagrams [2].

At present state of the ERC theory, of which the first version has been presented in [5], it is possible to show that there are about 130 tables of administrative information kinds. Their specification, as well as their structure and the structure of relationships between them, are here neglected because of editorial limits. The key attributes of these tables belong to the set of only 20 **structural attributes**. One of them is time and the others are integer numbers. Among sets of their values there are distinguished 12 **database dimensions**. The dimensions are such sets of attribute values that sets of values of the key attributes of all subclasses of information kinds are their subsets.

The dimensions of any ERC database are sets of values of the following information kinds attributes (Fig. 15.1):

s   item number of widely comprehended organizational systems, $s \in S$,
b   item number of basic control objects in a given elementary organizational system, $(s, b) \in SB \subset S \times B$,
k   item number of widely comprehended transitions in a given organizational system, $(s, k) \in TR \subset S \times K$,
o   item number of widely comprehended operation kinds, $o \in O$,
m   item number of widely comprehended information places, $m \in M$,
r   item number of widely comprehended resource kinds, $r \in R$,
e   specimen number of a given resource kind, $(r, e) \in RE \subset R \times E$,
q   item number of resource parameters, $q \in Q$,
h   item number of time scales and organizational levels, $h \in H$,
t   initial time of a sample period in a given time scale, $(h, t) \in HT \subset H \times T$,
n   item number of facts concerning either given subsystems or operations, activities or transitions located inside of these subsystems, $n \in N$, e.g. item number of execution orders of a given located operation (s, o): $(s, o, n) \in SON \subset SO \times N$,
y   item number of data processing phases, $y \in Y$.

The other structural attributes are item numbers of aggregated entities, containing other entities which are indexed by the attributes belonging to the same dimensions:

u   item number of resource categories, $u \in U \subset R$,
l   item number of resource lots, $(r, l) \in RL \subset RE, L \subset E$,
c   item number of resource aggregation counts, $c \in C \subset M$,
w   item number of work subsystems, $w \in W \subset S$,
j   item number of transition units in a given organizational system, $(s, j) \in TJ \subset TR, J \subset K$,
g   item number of operations groups, $g \in G \subset O$,
on   item number of process kinds, i.e. superior operation kinds, $on \in ON \subset O$,

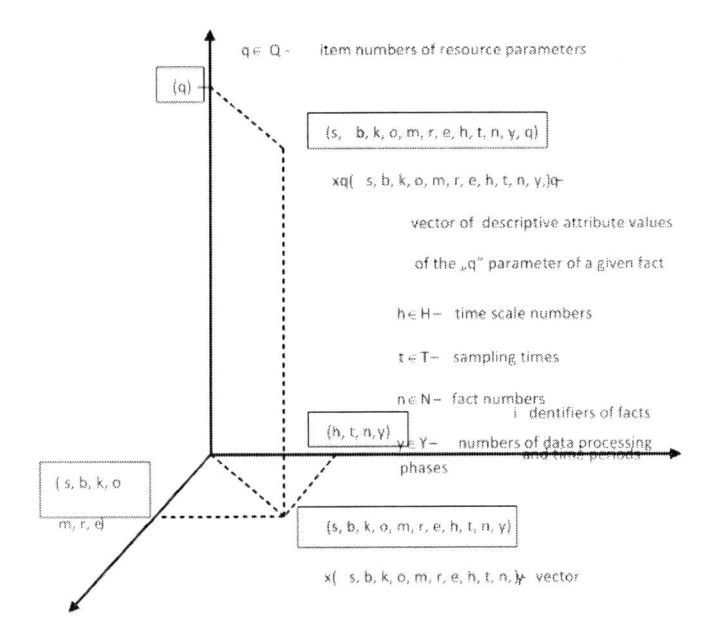

identifiers of active and passive entities

$s \in S$ - item numbers of organizational systems
$b \in B$ - item numbers of basic control objects in a given elementary system „s"
$k \in K$ - item numbers of transitions in a given organizational system „s"
$o \in O$ - item numbers of operation kinds
$m \in M$ - item numbers of information places
$r \in R$ - item numbers of resource kinds

Fig. 15.1 The demonstrative diagram of the data cube structure in the framework ERC system

nn   item number of aggregated facts, $nn \in NN \subset N$, e.g. item number of execution orders of a given located process (sn(s), on, nn) containing execution orders of stage operations (s, o, n): $(s, o, n, on, nn) \in SONON \subset S \times O \times N \times ON \times NN$.

The most important subclasses of the ERC database dimensions are sets of item numbers of operation and resource kinds:

a   item number of elementary activity kinds, $a \in A \subset O$,
v   item number of procedures in the function library of an ERC system, $v \in V \subset O$,
os   item number of system operation kinds, $os \in OS \subset O$,
op   item number of productive operation kinds, $op \in OP \subset OS \subset O$,
oh   item number of preparative operation kinds, $oh \in OH \subset OS \subset O$,
oi   item number of administrative operation kinds, $oi \in OI \subset OS \subset O$,
d   item number of number of productive information kinds, $d \in D \subset R$,

f   item number of financial resource kinds, $f \in F \subset R$,
i   item number of administrative information kinds, $i \in I \subset R$,
p   item number of consumable resource kinds, $p \in P \subset R$,
z   item number of reusable resource kinds, $z \in Z \subset R \subset S$.

15.2 Organizational Information-Transition Nets

Software of planning and control systems, as for all computerized information systems, is designed to data processing. Thus a design of such a system should precisely determine:

- data structures,
- functions and processes of data processing,
- organization of data processing systems.

Information flow between data processing procedures in present planning and control systems is not a direct one. Information is recorded to proper areas of a database and/or to registers of measurement, control and execution devices and read later on from them. In the ERC theory sites of information memory and sites of information processing are called correspondingly **information places** and **transitions**.

Each information place $m \in M$ is one of disjunctive subsets of the administrative information elements $(i, e) \in IE \subset RE \subset R \times E$. Each transition $(sn, k) \in TR \subset S \times K$ has got its own transition procedure $v(sn, k) \in V \subset O$. Transition executions are database transactions. Transition procedures may be remembered as stored procedures of the ERC database.

For modelling ERC systems **Organizational Information-Transition Nets** (OITN) have been created [5]. In OITN, likewise as in Coloured Petri Nets (CPN) [3], information places are presented as ovals and transitions as rectangles (Fig. 15.2). Pages of a hierarchical CPN correspond to organizational systems belonging to multilevel OITN. OITN places, which are adjacent to a given transition, are sets of such tuples from database tables that are input or output data of this transition. The most important differences between OITN and CPN are following:

- There is exactly one token in each OITN place and therefore tokens are not shown in OITN diagrams.
- Each OITN arc corresponds to two CPN arcs, which represent reading from and writing to an information place, i.e. to a proper ERC database area.
- In each OITN place data structure is the same as in relational databases. So it is much simpler than for CPN ML and much easier for database implementations.
- Transition procedures may be written in any SQL dialect. Thus they are much simpler and easier for implementation than code segments written in CPN ML.

Parallel transitions, which have common sets of input and output information places, are grouped into **transition units** $(sn, k) \in TJ \subset TR$ In the framework ERC system all transition units (without some exceptions) occur in pairs. For each decision unit

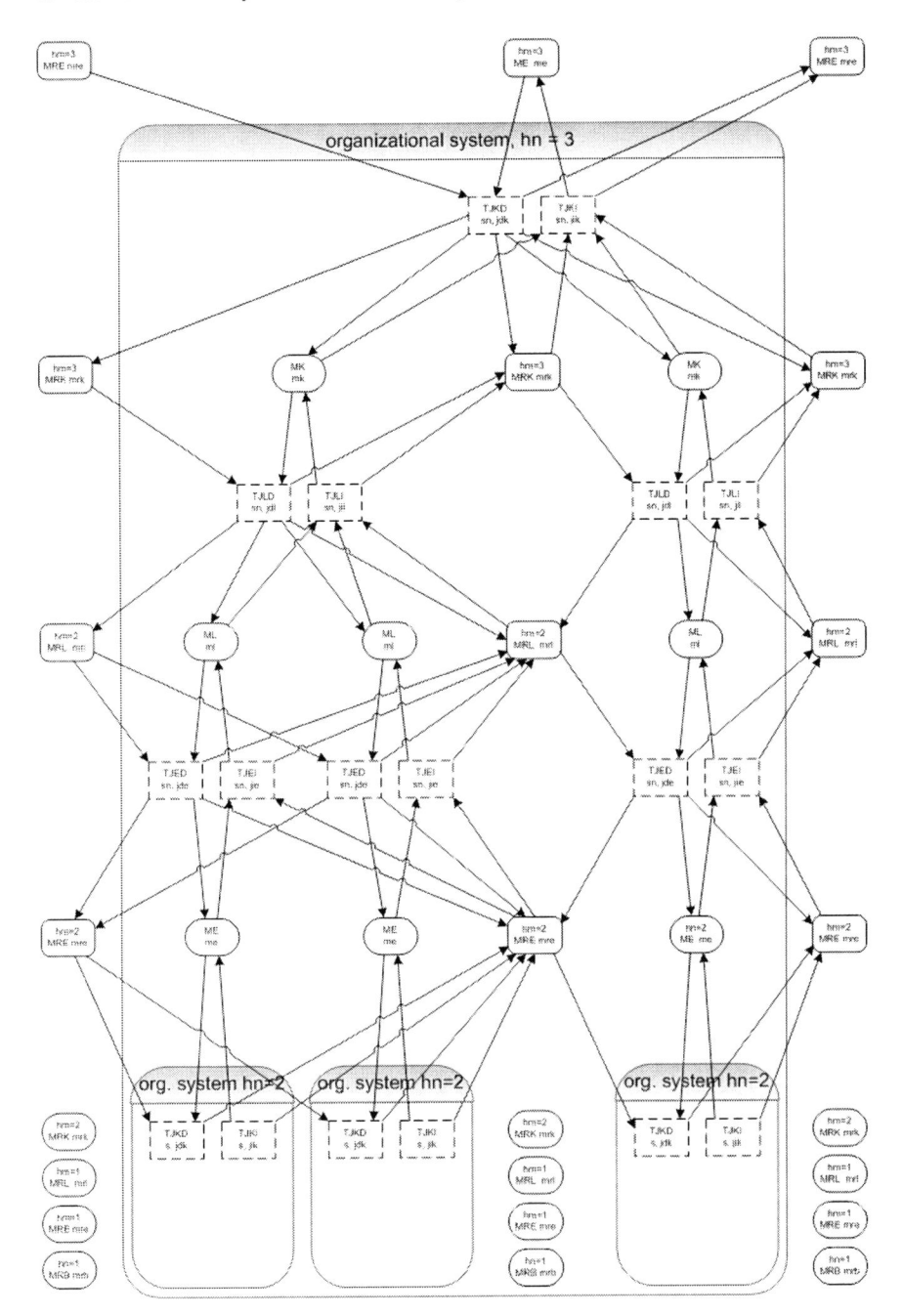

Fig. 15.2 The OITN model of coordination, allocation and execution planning in an organizational system

processing general decisions and available information into more detailed decisions, there is an information unit, that processes detailed information into aggregated general information, which are needed for transitions belonging to higher functional layers and to higher organizational levels.

15.3 Organizational Structure

A **located process** (sn, on) $\in$ SNO $\subset$ SN $\times$ ON $\subset$ S $\times$ O, situated in a given organizational system sn $\in$ S is an ordered set of **located operations** (s, o) $\in$ SO $\subset$ S $\times$ O, situated in its subsystems s $\in$ S, and **located resources** (m, r) $\in$ RMB $\subset$ MRB $\times$ O, which are situated in **basic places** m $\in$ MRB $\subset$ M. Located resources separate stage located operations of a given located process. **Resources** r $\in$ R = P $\cup$ Z $\cup$ F $\cup$ D $\cup$ I are classified into consumable (material products), reusable, financial, informational (productive information which is processed for customer needs) and administrative ones. **Operations**, as subprocesses, are classified into productive, preparatory and administrative ones. Transition procedures, kinds of elementary activities and groups of operations also belong to widely comprehended operations o $\in$ O = OP $\cup$ OH $\cup$ OI $\cup$ V $\cup$ A $\cup$ G. All processes in a given enterprise, except for **primary business processes**, are stage operations in processes of higher level on $\in$ ON $\subset$ O. Primary business processes are executed in a **primary organizational system**, containing the enterprise plants, plants of its customers and suppliers and other elements of its environment. **Elementary processes** (situated in elementary organizational systems) contain elementary activities.

An **organizational system** sn $\in$ S is the part of an ERC system, which is designed for executing a determined set of processes on $\in$ ON. Analogously, each **organizational subsystem** s $\in$ S belonging to the given system sn(s) is designed to execute a determined set of operations o $\in$ O, which are stage operations of processes located in this system. **Organizational level** is a set of organizational systems situated in the same level of the enterprise hierarchy tree. Typically an ERC system has four organizational levels:

h=4   primary organizational system,
h=3   production plants,
h=2   organizational cells,
h=1   workstations, i.e. elementary systems, which do not have their own organizational subsystems.

In **elementary organizational systems** (h=1) there are no subsystems. **Basic control objects** (sn, b) $\in$ SB $\subset$ S $\times$ B, that belong to them, are pairs of basic (decision and information) transition units. **Basic decision transitions** do not generate any decisions, but only receive decisions concerning basic objects. **Basic information transitions** do not process any information, but only emit **reports** (e.g. as measurement results) about executing decisions, which were previously received by decision transitions. Each basic object (sn, b) $\in$ SB $\subset$ S $\times$ B has got its own elementary activity a(sn, b) $\in$ A $\subset$ O. Its executions begin from decision transition executions and finish (always after finite time period) with information transition executions.

Apart from organizational subsystems s $\in$ S there are **work subsystems** w $\in$ W $\subset$ S in organizational systems. They are arrangements of parallel organizational subsystems with input or output products located in common sets of border places. Organizational level of work subsystems are not higher, but the same as for their component organizational subsystems. In a typical ERC system they are:

h=3   enterprise, composed of one or many production plants,
h=2   department, composed of one or many organizational cells,
h=1   workcentre, composed of one or many workstations.

15.4 Functional Structure

The OITN structure, represented by information places, transitions and connecting them arcs, univocally models not only organization of data processing in an enterprise, but also organization of this enterprise (Fig. 15.2). Each organizational system has one pair of coordinative transition units. Each work subsystem has one pair of allocation planning transition units and one pair of reengineering transition units (which are not shown in the simplified diagram from Fig. 15.2), while each organizational subsystem has one pair of executive transition units. So transition units form natural **functional layers** of coordination, reengineering, allocation and execution planning (Fig. 15.3). Information places, which are located between them, are laid in **informational layers** with just the same easiness. In a given organizational system the executive informational layer connects transitions from executive functional layer with coordinative transitions inside of its organizational subsystems. This is a model of hierarchical relationships between organizational systems from different organizational levels.

Informational layers are situated directly above functional layers, whose time scales are the same, but their decisions may origin from each of higher functional layer. Analogously, their information may be directed to each one of them. Each layer has its own **time scale**, which is determined by length of time periods, for which decisions are made and reports are generated. For ERC systems it is assumed that total number of time scales and organizational levels is the same. For three higher layers of a given level the time scale is the same, while for the executive layer reports and decisions are issued with shorter time periods that correspond to the lower organizational level. It is one of reasons for greater precision of executive decisions than for corresponding allocation planning decisions.

In an elementary organizational system there is only three functional and two informational layers. Basic transition units form the layer of receiving control signals and issuing measurement signals of basic control objects. Controllers of these objects belong to the direct control layer. The coordination layer in a given elementary system is the same as for higher level systems. It is composed of two transition units of coordination control.

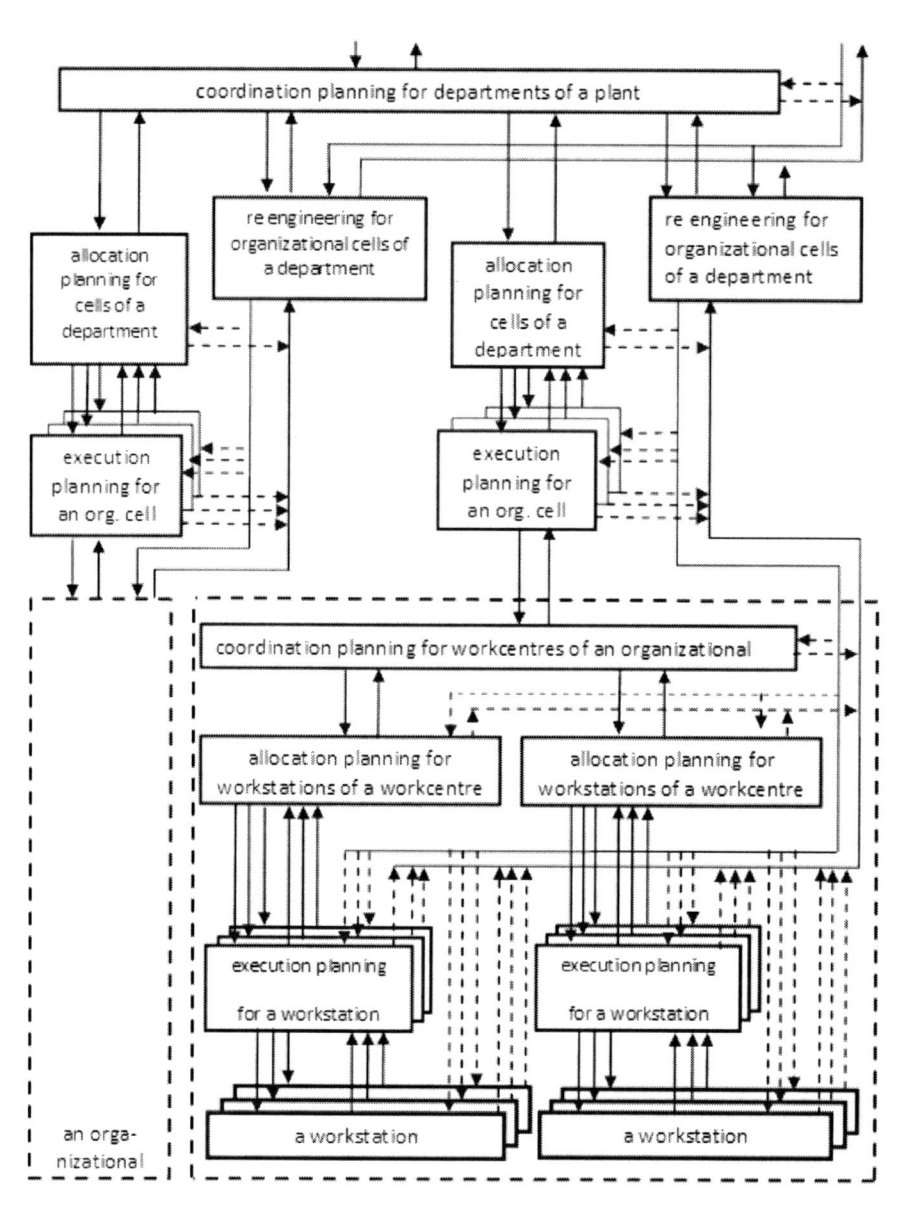

Fig. 15.3 Functional layers in exemplary organizational systems of adjacent organizational levels

15.5 Conclusions

The ERC theory is not only generalization of properties noticed in some real computerized information systems. It is deduced from general structural features of processes running in enterprises and from generally formulated tasks of integrated planning and control systems. It encourages to the statement that every system of ERP, MES or SCADA, irrespective of the trade and size of the enterprise in which it is implemented, may be transformed into the framework ERC system with all its functions and data [5]. In spite of such declared generality of the ERC theory the structure of the framework ERC system is relatively simple and univocally determined.

Generality of the ERC theory may not be formally proven but it may be justified by demonstrating its usefulness for real management and control systems and by its comparisons with MRP II [4] and ISA-95 [1] standards.

References

1. ANSI/ISA-95: Enterprise-Control System Integration. Part 1,2,3,5 (2000-2007)
2. Beynon-Davis, P.: Database Systems. Macmillan Press Ltd., Basingstoke (2000)
3. Jensen, K.: Coloured Petri Nets. Springer, Berlin (1997)
4. Landvater, D.V., Gray, C.D.: MRP II Standard System. Oliver Wight Publications (1989)
5. Zaborowski, M.: The Follow-up Enterprise Resource Control. Ed. PK J. Skalmierski, Gliwice (2008) (in Polish)

Chapter 16
Implementation of Changes to the System in the Model of Multiple Suppliers vs. the Model of a Single Supplier

Wojciech Filipowski and Marcin Caban

Abstract. In today's information society, based on easy access to the means of communication, in order to maintain competitiveness on the market, it is necessary that the market participants implement new technological solutions. In case of Corporation, there is a need to diversify the suppliers of IT solutions. Competition between suppliers is now seen as a desirable way of stimulating the development of technology and improving the quality of services, while providing opportunities to optimize the investment costs. This article presents the situation in a particular area in which on one hand we are dealing with one supplier, and on the other hand allow to confront the suppliers in the market fight for contract for the implementation of changes to the system functioning as an integrated system environment. Against this background, there are technical differences shown in the change implementation process, as well as aspects related to the choice of the implementation model.

16.1 The Process of Implementing Changes to the System

The process of implementing changes to the system in the Organization (company, corporation) can be divided into the stages shown in Fig. 16.1.

This model of implementation is one of the options acceptable, but the steps listed in this example, are sufficiently representative that can be used to bring this particular case, and to draw more general conclusions.

Wojciech Filipowski · Marcin Caban
Silesian University of Technology, Faculty of Automatic Control,
Electronics and Computer Science, Akademicka 16, 44-100 Gliwice
e-mail: Wojciech.Filipowski@polsl.pl

E. Tkacz and A. Kapczynski (Eds.): Internet – Technical Development and Appl., AISC 64, pp. 141–148.

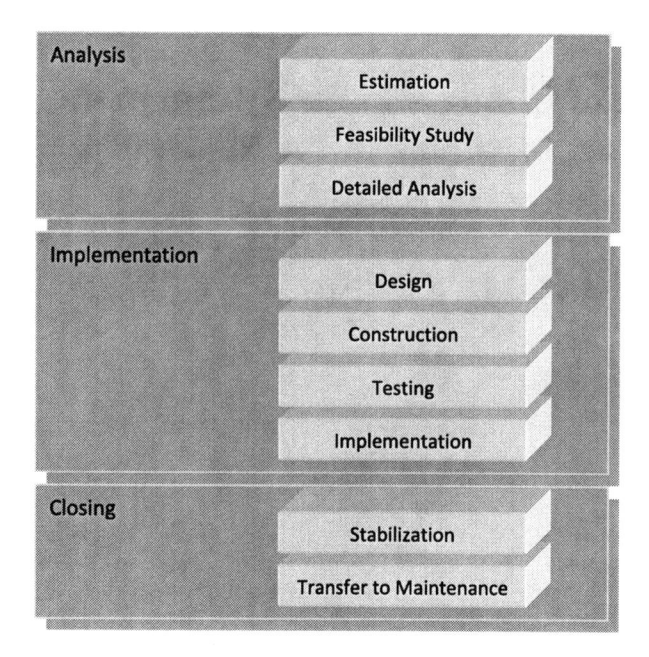

Fig. 16.1 Diagram of the implementation of changes to the system

16.2 The Lines Dividing the Process – The Responsibility of Suppliers and Points of Contact

The following provides diagrams of the stretch of the change implementation process to the system in the Organization. The process has been presented here in terms of participation in the supplier or suppliers. In this perspective potential contact points of suppliers are very clearly shown on the background of the main stages of the implementation process. These points of contact are boundaries of suppliers' responsibilities, on a certain stage of the process, designated by sections separating blocks of the implementation process diagram. On each side of a boundary the responsibility for the implementation can be taken by a different supplier. Here are the main categories of division of the change implementation process in an information system.

Note! Blocks of the diagram - showing the stages of implementation process - described by a letter underlined mean the sole or the predominant responsibility of the supplier, while the rest are areas in which the supplier provides support to the Organization, and the predominant or exclusive responsibility falls on the members (employees) of the Organization.

Figure 16.2 shows the cross-process variant of implementation. The variant represents the change in the information system implemented as a single implementation project, while the various stages of implementation are subjects to division of responsibility. These stages somewhat cut the whole process across its course.

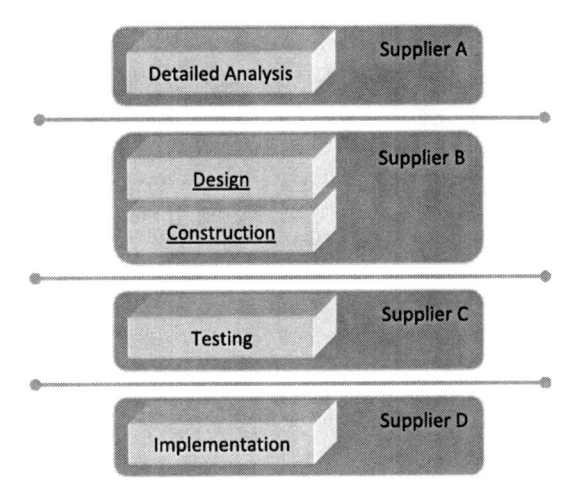

Fig. 16.2 Cross-process implementation

This situation allows the separation of a part of the project and transfer it to the implementation to a supplier, while the remaining stages may be carried out either by another supplier, or by the Organization with the support of a supplier. Typical in this case is that the process of implementation is sequential, i.e. results of previous stages are passed along to the liability of those responsible for subsequent process steps.

It should be noted that in a particular case, we may have to deal with just one supplier, which, in Design and Production stages takes over the responsibility for implementation from the Organization, and then transmits the products back to the Organization to continue implementation.

The longitudinal variant (Fig. 16.3 represents a complex change in the implemented system. This is an implementation process of many flows - streams. The change in this case consists of streams carried out in parallel and synchronized in time, while each of concurrent streams may be of different supplier responsibility. Variant of a longitudinal model of the above is a change that is not synchronized in time, but in such a situation we do not treat it as a complex change, but as many smaller changes carried out separately. Turning back to the situation shown in the diagram, the longitudinal model supports multithreaded project, which Detailed Analysis, Testing, and Implementation stages are treated in the project plan as individual tasks within the Organization.

A characteristic feature of this variant is that in this case suppliers do not provide each other with the results of the work at various stages of implementation, and the project as whole is integrated by the Organization.

One should be aware, however, that such variant is possible to apply in the real life only to the Organization, which has very strong competencies in the implementation of solutions, especially in the management of source code. Among other things, because of the source code management problem that is crucial to solve, the option is not used alone but as part of a mixed model, which was raised later in this paper.

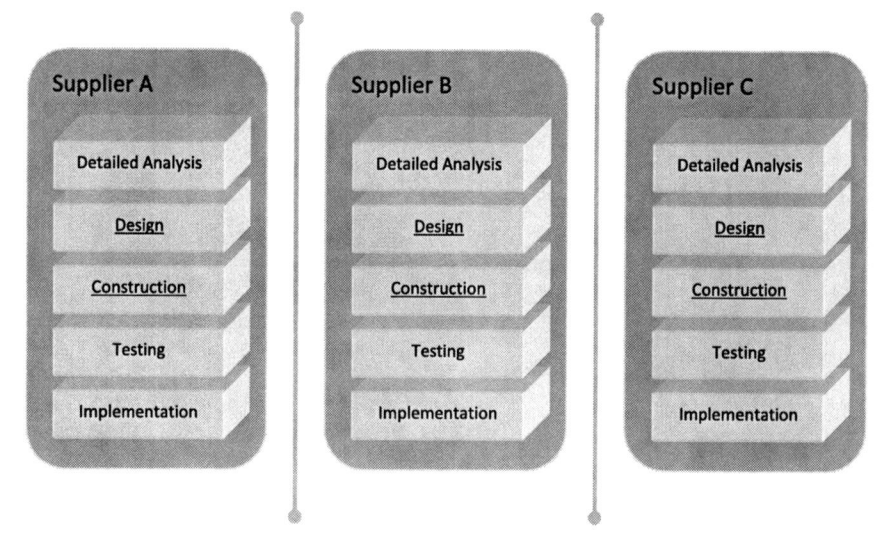

Fig. 16.3 Longitudinal variant of implementation process

16.3 The Specificity of the Single Supplier Model

Single supplier model can be represented by the shown in Fig. 16.4 diagram of the process of implementation.

Fig. 16.4 Single supplier
model

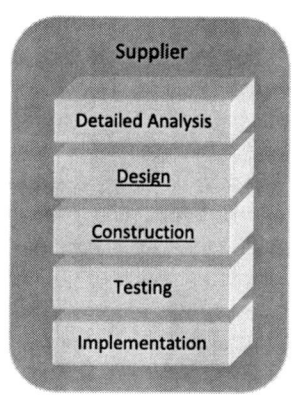

The model of single supplier is the most classic example of the implementation process in which there is only one supplier and during the implementation it does not interacts with other suppliers. This situation does not change the fact that the supplier is working together with the Organization at which the implementation is carried out, resulting in the movement of responsibility between the supplier and the Organization. The provider in this model is present as an independent external party.

16.4 The Specificity of the Multi-supplier Model

The next diagram (Fig. 16.5 shows the hybrid model - a characteristic of the model with multiple suppliers.

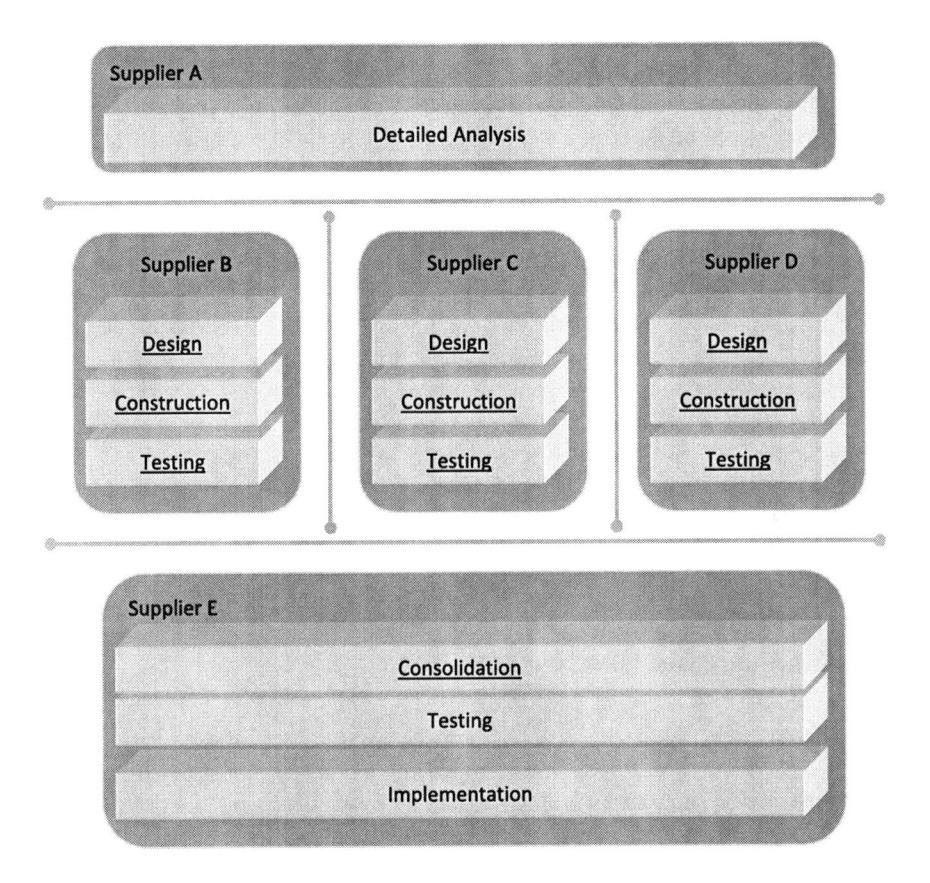

Fig. 16.5 The multi-supplier model

The model of the process of implementing shown in the above scheme is the most complex of the presented and illustrates the division of responsibilities and interactions between suppliers, both along the course of the implementation process, as well as on concurrent basis.

The process of implementation is partly parallel and relies on using multiple suppliers in concurrent implementation stages. Because of the role the suppliers play in the process, those supplier are called Development Suppliers. In the process, each development supplier is responsible for his production stream, and then under the control of the Organization shall transmit the products of their work to another supplier sequentially along the process.

It is worth mentioning here an additional, yet not discussed stage - Consolidation.

This is an essential element to integrate all the fragmentary changes to the system submitted by the Development Suppliers. Because of the functions realized in the process of implementation, the supplier responsible for this stage is often referred to as a Solution Integrator. Among the tasks in the main Integrator responsibility are: authorization of the source code, source code merge together with the removal of source code conflicts, execution or support of regression testing, the merged code transmission to the production system.

An important element of the model are tests.

They are held at several implementation stages and shall be implemented by all suppliers, and are subjects to optimization due to their partial concurrency.

Each of the providers in the ongoing development of their stream - test the implemented partial change to the system. Integrator Supplier after the code merge performs the tests in the course of his responsibility - including authorization, and supports testing carried out by the Organization.

Such a diversity of tests performed during the implementation allows for effective detection of errors in the software, which helps the best quality of the final product which is a system or a new version of the system.

With regard to ensuring a comprehensive management system source code, in large organizations and implementing projects involving multiple suppliers, this model is used in practice as the most versatile - allowing for the optimization of expenses through concurrent or partially concurrent execution of the most costly stages of implementation.

16.5 Summary and Conclusions

Organization in support of its operational activities by means of information systems facing the need to change these systems, together with its own development and due to changing market requirements. Each of the implementation models shown in previous chapters, because of its specificity, has a legitimate use of the area, each also can cause or reduce risks and limits to the Organization. Therefore, when deciding on the strategy of system development the Organization should be aware of these circumstances to secure the development of the system with the optimal cost in the long term.

The following analysis presents a model of the system development with one supplier, with many suppliers and their application areas.

The single supplier model is an easily managed one within an Organization. In this model the Organization responsibility comes down to quality and progress control of the supplier, and the process of implementation is based on the methodology used by the organization. This is checked in the case of relatively small complexity of changes or low frequency of occurrence, or in the case when, on account of on technical considerations, such as a unique technology - the market does not have enough diversity of suppliers.

In the long run, however, this model for the implementation of the changes could lead to monopolize the area by the system supplier, which introduces the risk of gradually increasing the cost of change implementation by the absence of competition between suppliers. Short-term benefit can be relatively small cost of necessary infrastructure needed by the Organization to provide as test environment or internal resources to support the management of the project an the supplier.

The multiple supplier model places relatively large coating on the Organization in terms of the need for a formal regulation of relationships with multiple suppliers. Often there is a need for considerable investment in test system environment infrastructure, as well as the repository of source code infrastructure and support, or system development environments to be shared to suppliers if needed.

So not in every case the model of multiple suppliers seems optimal. But from the Organization perspective of developing dynamically in large scale, in the long term, the multi-supplier model appears to deliver measurable benefits. These include:

- in the case of complex change there is a possibility to carry out certain stages of implementation concurrently, such as Production, or Testing;
- by transferring products between suppliers and introducing code authorization procedure, there is a better guarantee of the quality of work of individual suppliers;
- an option for the Organization to exchange suppliers in subsequent editions of the changes to the system, so many providers have access to knowledge about the system and thus is likely to allow real competition in the area;
- potential chance for large changes cost reduction through the competition of suppliers.

Observing the dynamic nature of large corporate organizations one can see that they attach great importance to the costs associated with investments in the development of information systems. Those systems help the information society an Organizations to develop and keep trying to find the best strategy in this area, which offers the best return on capital.

At present, it appears that the information systems development model in Organizations, using multiple suppliers gives optimal results. However it is important to pay attention to the role of the Integrator Supplier, which gradually increased its role, which in some cases, can guarantee the supplier obtain the dominant position in comparison to other competing suppliers. From this observation comes the conclusion that currently available methods and models of implementation continue to develop and will be subject of further evolution, in response to emerging needs of the Organization, to provide their services to.

Perhaps this development will have be adjusted to the needs of the Organization other than ever before due to the global financial crisis, but one can assume with a high degree of probability that the development of methods for the implementation of changes in information systems will continue to reflect the real needs of the market.

References

1. Project Management Institute Inc., A Guide to Project Management Body of Knowledge, PM-BOK Guide-Fourth Edition (2008)
2. Fred Brooks, The Mythical Man-Month: Essays on Software Engineering (1995)
3. Yourdon, E.: Death March: The Complete Software Developer's Guide to Surviving "Mission Impossible" Projects (1999)
4. Berkun, S.: The Art of Project Management (2005)

Chapter 17
Standardization of Software Size Measurement

Beata Czarnacka-Chrobot

Abstract. The measurement of software products is an area of software engineering, which cannot be considered as sufficiently mature not only in terms of practice, but also in terms of knowledge maturity. That is why over the last couple of years significant intensification of works could have been observed, which aimed to standardize the best practices of software products measurement. Various ISO/IEC norms have been developed as a result of these works, filling an important gap in the software engineering. One of the most important groups of ISO/IEC standards concerns the software product size measurement. Normalization of such measurement is mainly aimed at reducing unnecessary diversity in the area of software size measures, ensuring compatibility between the standardized approaches as well as their usefulness, especially for Business Software Systems. Thus the aim of the paper is to provide a synthetic presentation of these very aspects of the software products size measurement standardization.

17.1 Introduction

Definition of software engineering adopted by the Institute of Electrical and Electronics Engineers (IEEE) reads that it is *"the application of a systematic,disciplined, quantifiable approach to the development, operation, and maintenance of software"* [4]. Quantifiable approach means that the measurement of software processes and products should constitute immanent feature of this knowledge and life discipline. Yet software engineering cannot boast of high maturity level with regard to the measurement of what makes its subject, including most of all software size. Meanwhile, *"measurement of software size (...) is as important to a software professional*

Beata Czarnacka-Chrobot
Faculty of Business Informatics, Warsaw School of Economics,
Al. Niepodleglosci 164, 022-554 Warszawa, Poland
e-mail: bczarn@sgh.waw.pl

E. Tkacz and A. Kapczynski (Eds.): Internet – Technical Development and Appl., AISC 64, pp. 149–156.
springerlink.com

as measurement of a building (...) is to a building contractor. All other derived data,including effort to deliver a software project, delivery schedule, and cost of the project, are based on one of its major input elements: software size" [17].

Hence the right measure of software size has been sought out for several decades now. These efforts have brought fruits only recently: one of the concepts of software size measurement along with several methods based on that concept were normalized by the ISO (International Organization for Standardization) and IEC (International Electrotechnical Commission). Many years' verification of various approaches showed that what for now deserves standardization is just the concept of software size measurement based on its functionality, being an attribute of first priority to the user. Normalization of the concept and methods of such measurement is mainly aimed at reducing unnecessary diversity in the area of software size measures, ensuring compatibility between the accepted approaches and relevant rules as well as ensuring their usefulness, especially for Business Software Systems (BSS), because:

- BSS are one of the fundamental IT application areas;
- BSS development/enhancement projects often constitutes serious investment;
- In practice, ready-made BSS rarely happen to be tailored to the particular client business requirements therefore their customisation appears vital;
- Rational ex ante and ex post valuation of unique (at least partially) BSS, being of key significance to clients, encounters serious problems in practice;
- From the provider's perspective, the BSS development/enhancement projects are particularly difficult in terms of management, which basically results in low effectiveness as compared to other types of software projects [16] [18].

17.2 Software Size Measures

Basic approaches to the software product size measurement may be reduced to perceiving it from the perspective of:

- Length of programmes, measured by the number of the so-called programming (volume) units. These units most of all include Source Lines of Code (SLOC), but number of commands, number of machine language instructions are also taken into account. However, these units measure neither size of the programmes nor their complexity but only the attribute of "programme length" yet thus far these are them that in practice have been employed most often with regard to the software size [2] [17].
- Software construction complexity, measured in the so-called construction complexity units. Most of hundreds of such measures having been proposed are limited to the programme code [2] yet currently these units are used mainly in the form of object points [17]. These points are assigned to the construction elements of software (screens, reports, software modules) depending on the level of their complexity.

• Functionality of software product, expressed in the so-called functionality units. They most of all include function points, but also variants based on them such as: full function points, feature points, or use case points. These points are assigned to the functional elements of software (functions and data needed to complete them) depending on the level of their complexity, not to the construction elements as it was the case of object points.

Synthetic comparison of various software size measures against a background of key requirements set for such measures were presented on Fig. 17.1 (to a much broader extent this issue was discussed in [1]). Details displayed therein clearly indicate the reasons why functionality units were recognised as the most appropriate measure of software size not only by ISO/IEC but also, among others, by Gartner Group [3] as well as by International Benchmarking Standards Group (ISBSG) [5]. They show no limits being characteristic of programming units and construction complexity units, although one may have reservations as to their versatility and relatively high complexity of the methods based on them. However, it is hard to expect that the method of measurement of software products, being by nature complicated, would be effective yet simple.

Requirement towards measure	Programming units	Construction complexity units	Functionality units
Unequivocalness of definition	Freedom in formulating definitions (differences as big as even 5:1)	Depending on the method	In methods normalized by ISO/IEC
Possibility to make reliable prognosis on the size relatively early in the life cycle	Possibility to calculate programme length only for the existing code	None – with regard to programming units and object points	As early as at the stage of requirements specification [5]
Base for the reliable evaluation of the all phases work effort	Final programme length does not fully reflect the whole work done	Final software size does not fully reflect the whole work done	Relatively high reliability as early as at the stage of requirements specification [5]
Software size being independent of the technology employed	Programme length determined by the language employed	Size being dependent on the technology employed	Size depends on functional user requirements
Possibility to compare software written in different languages	Lack of such direct possibility	Lack of such direct possibility	Size doesn't depend on language used
Measuring size in units being of significance to a client/user	No significance to a client/user	Secondary significance to a client/user	Measurement from the point of view of functionality
Possibility to compare delivered size vs. required size	Inability to make reliable prognosis	Inability to make reliable prognosis	Thanks to the possibility of making reliable prognosis
Possibility to measure all software categories	Yes	Depending on the method	Depending on the method
Easiness of use	Yes	No	No

Fig. 17.1 Synthetic comparison of software size measures

17.3 Standardization of Software Functional Size Measurement Concept

Set of rules regarding software size measurement in functionality units was included to the six-part **ISO/IEC 14143** norm [7]. For such measurement this standard proposed definition of functional size, which is understood as *"size of the software derived by quantifying the Functional User Requirements"*. While Functional User Requirements (FUR) stand for the *"sub-set of the User Requirements describing what the software does, in terms of tasks and services"*. Hence functional requirements in this norm, due to their importance and need to ensure objectivism of measurement, are treated disjointly when combined with other requirements of non-functional character. The elementary unit of FUR defined in the ISO/IEC 14143 standard and used for measurement purposes is called Base Functional Component (BFC). Thus this standard normalizes the concept of the so-called **Functional Size Measurement (FSM)** of software products.

First, definitional part of the ISO/IEC 14143 norm constitutes conceptual basis for such measurement and therefore provides base for the remaining parts of this standard, among which are the following:

• Part 2, containing the rules of assessing conformance of the potential Functional Size Measurement Method (FSMM) with definitional part - FSMM is defined as a specific implementation of FSM concept defined by a set of rules, which conforms to the mandatory features (definitions, characteristics and requirements) of such measurement [7].

• Part 3, describing processes designed for objective and consistent verification of various aspects of FSMM (e.g. team competences, employed procedure) in a way that allows users to choose method which is best tailored to their needs.

• Part 4, comprising set of references designed for FSMM verification, divided into BSS, real time systems and scientific software, being useful when comparing the measurement results achieved with the use of different FSMM.

• Part 5, featuring description of software classes (functional domains), for which possibility of using FSMM is declared, as well as providing characteristics of the such domain defining process on the basis of the given set of FUR.

• Part 6, comprising the rules of selecting, among methods recognised by the ISO/IEC, the FSMM which would be suitable for given domain as well as giving guidance on how to use FSM concept to support software development, enhancement and maintenance projects management.

According to the ISO/IEC 14143 norm the process of using FSMM should comprise the following steps: defining the scope of functional size measurement, identifying the FUR contained within the scope of functional size measurement, identifying the BFC contained within the FUR, classifying the BFC with regard to their type, assigning appropriate value to each BFC, and calculating functional size.

After about 30 years of improving various software FSM techniques five of them (out of over 20) have been acknowledged by the ISO/IEC as conforming to the rules laid down in the ISO/IEC 14143 norm and have taken on the form of the following standards:

- **ISO/IEC 20926** [11], in which was approved the function point method in the version developed by the International Function Point Users Group (IFPUG)
- **ISO/IEC 20968** [12], describing the function point analysis developed by the United Kingdom Software Metrics Association (UKSMA)
- **ISO/IEC 24570** [13], in which was approved the FSM method in the version proposed by the Netherlands Software Metrics Association (NESMA)
- **ISO/IEC 19761** [10], normalizing the method of FSM developed by the Common Software Measurement International Consortium (COSMIC)
- **ISO/IEC 29881** [14], in which was normalized the FSMM developed by the Finnish Software Metrics Association (FiSMA).

The FSM methods accepted by the ISO/IEC differ in terms of software measurement capabilities with regard to different functional domains (software classes). Therefore prior to choosing given method one should firstly evaluate its adequacy to the type of software, whose functional size is going to be measured.

17.4 Standardization of Software Functional Size Measurement Process

The ISO/IEC 14143 norm adhere to the **ISO/IEC 15939** standard [9], determining general rules and procedures for the software measurement process in compliance with the **ISO/IEC 15288** norm [8] which, on the other hand, defines processes of the system's life cycle.

One of the steps of the size measurement process defined in the ISO/IEC 15939 standard is procedure of selecting a method that will be used to measure its size. According to this procedure, selection of FSM method being best tailored to the user's needs should consist of the following activities:

1. Characteristics of organisational units of software user with regard to the measurement process.
2. Identification of their information needs towards measurement process.
3. Selection of appropriate FSM method on the basis of prospective methods identification.

Requirements towards appropriate FSM method vary depending on the organization's character. For example, financial institutions usually choose the method, which correctly measures the BSS while chemical company, by reason of its basic activity, would rather require measurement method being suitable for the real-time systems. Thus selection of appropriate method should begin with dividing organisation's software into functional domains. Choosing method adequate to the needs would also depend on how its result is planned to be used. If an organisation intends to use the measurement results also for the purpose of comparing its productivity against industry data, it is recommended to choose the method being relatively popular in the given industry, for which such data exist. In the case it only needs cursory, rough estimation of functional size, the requirements towards appropriate method of its measurement will get reduced.

Thus selection of the most adequate method may be facilitated by the following activities:

• Identification of organisation's areas which are to be measured and dividing their software into functional domains in compliance with the 14143 norm (Part 5) that distinguishes the following categories of such domains: business applications, real time systems, scientific software, and infrastructure software.

• Finding out how an organisation acquires software: whether it is developed inside organisation, or organisation purchases ready-made software without customisation, or it purchases ready-made software and tailors it to its needs, or software is being developed by an outside provider from scratch. This determines the scope of activities, which the selected method should take into account as well as the purpose of measurement process.

• Identification of software execution processes used in an organisation and being within the scope of activities, which the method is supposed to cover. Such identification should be based on processes defined in the **ISO/IEC 12207** norm [6] as processes which should be subjected to evaluation although it is possible to consider processes that are specific to an organisation. Therefore the processes may comprise: planning, ordering, development, implementation and management of the possessed software resources.

• Establishing organisational measurement procedure on the basis of recommendations featured in the ISO/IEC 15939 norm. Organisation's capabilities in the area of measurement affect such aspects being of significance to this process as e.g. accuracy of results, discipline in collecting information for measurement purposes, selection of individuals responsible for the measurement, required level of their competence and experience as well as time designed for this process.

Next step in selecting the right FSM method is identification of information needs towards measurement process. It is aimed to define and prioritise requirements towards such method for a specified product recipient. Identification of information needs should begin with analysis of purposes for which the measurement results will be used - while hierarchisation of these purposes is worth doing. Next the appropriate requirements regarding the results (e.g. level of repeatability, accuracy, conversion possibility) should be determined for each purpose.The ISO/IEC 14143 norm (Part 3 and/or 4) proves helpful here. As a result of this step are determined software basic functional components (BFC) that are to be measured, as well as stages of the project at which the results corresponding with these components should be achieved.

In the ISO/IEC 14143 norm it is stated, that:

• There are no functional domains constraints for the accepted part of the IFPUG and NESMA methods nor for the FiSMA method.

• The UKSMA method is adequate for any type of software provided that the so-called logical transactions may be identified in it (the rules were developed as intended for BSS). The rules support neither complex algorithms characteristic of scientific and engineering software nor the real-time systems.

• The COSMIC method is adequate for BSS, real-time systems and hybrid solutions combining the two. There are constraints for software with complex mathematical algorithms or with other specialised and complex rules (e.g. expert, simulation, self-learning, weather forecasting systems) and for software processing continuous variables (e.g. computer games or musical instruments software).

ISO and IEC allow for selecting method other than the methods approved by them yet they recommend that it conforms to definitional part of the ISO/IEC 14143 norm. According to the requirements of these organisations, after selecting the FSMM one should define its steps to be undertaken in the given case along with the method of collecting data which are to be measured, the way of communicating measurement results, storing, managing, and reviewing them which is supposed to promote standardization of measurement procedures. On the other hand, it should promote accuracy of the results achieved thanks to this measurement. It is also recommended to carry out measurement with the use of relevant supporting tools.

17.5 Concluding Remarks

Using formal approaches in the software engineering practice is of great significance to the software projects stakeholders. It helps build clients' trust in potential providers thus making it easier to choose product supplier following objective and reliable criteria. Thus, if software organizations want to gain competitive advantage, they should be motivated to implement the so-called good practices being included in such standards. They sort out such organizations' activities, rationalize planning, allow to control the key attributes on a current basis which makes it possible to accelerate the reaction to the undesired situations, they also allow to increase a chance to deliver products of respectable quality, to identify areas that are in need of improvement therefore reducing uncertainty of activity. Although implementation of formal approaches quite often requires considerable expenditure, it may at first introduce some constraint to the activity flexibility while certificates neither guarantee effective client's requirements fulfilment nor they are a necessary prerequisite for it - they, however, undoubtedly contribute to this strongly. This gets confirmed by the results of surveys, among others of Standish Group, according to which the use of formal approaches is one of significant factors of the effective software projects execution [15]. Thus implementation of formal approaches compliant with an organization business needs and goals may be regarded as an investment in the software processes and products improvement. However, every organisation must weigh benefits coming from the implementation of formal approaches against its costs, taking into account only those being truly useful in order to avoid excessive formalisation.

Usefulness of standardization of the software FSM concept and methods manifests in providing support through the norms, being synthetically presented in this paper:

- Compliance with other standards (e.g. ISO 9000)
- Initiatives on processes improvement (e.g. CMMI)
- Applying FSM concept in agreements (e.g. outsourcing ones)
- Effective universal comparisons
- Maturity of FSM process, which should be predictable, repeatable and controllable
- Identification of software functional domains
- Selection of the right FSM method
- Appropriate use of FSM methods (e.g. in tools)
- Development of alternative FSM methods.

References

1. Czarnacka-Chrobot, B., Kobylinski, A.: Ocena miar zakresu produktu programowego (Software Size Measures Assessment). Prace Naukowe Uniwersytetu Ekonomicznego we Wroclawiu, Wroclaw (in press, 2009)
2. Fenton, N.E.: Zapewnienie jakosci i metryki oprogramowania (Ensuring software quality and software metrics). In: Gorski, J. (ed.) Inzynieria oprogramowania w projekcie informatycznym (Software Engineering in IT Project), 2nd edn. Mikom, Warsaw (2000)
3. Gartner Research: Function Points Can help Measure Application Size. Research Notes SPA-18-0878 (2002)
4. IEEE Std 610.12-1990: IEEE Standard Glossary of Software Engineering Terminology. The Institute of Electrical and Electronics Engineers, New York (1990)
5. ISBSG: The ISBSG Report: Software Project Estimates – How accurate are they? International Benchmarking Standards Group, Hawthorn VIC, Australia (2005)
6. ISO/IEC 12207:2008 Systems and software engineering – Software life cycle processes. ISO, Geneva (2008)
7. ISO/IEC 14143 Information Technology – Software measurement – Functional size measurement – Part 1-6. ISO, Geneva (1998-2007)
8. ISO/IEC 15288:2008 Systems and software engineering – System life cycle processes. ISO, Geneva (2008)
9. ISO/IEC 15939:2007 Systems and software engineering - Measurement process. ISO, Geneva (2007)
10. ISO/IEC 19761:2003 Software engineering – COSMIC-FFP – A functional size measurement method. ISO, Geneva (2003)
11. ISO/IEC 20926:2003 Software engineering - IFPUG 4.1 Unadjusted functional size measurement method - Counting practices manual. ISO, Geneva (2003)
12. ISO/IEC 20968:2002 Software engineering - Mk II Function Point Analysis - Counting practices manual. ISO, Geneva (2002)
13. ISO/IEC 24570:2005 Software engineering – NESMA functional size measurement method version 2.1 - Definitions and counting guidelines for the application of Function Point Analysis. ISO, Geneva (2005)
14. ISO/IEC 29881:2008 Information Technology – Software and systems engineering – FiSMA 1.1 functional size measurement method. ISO, Geneva (2008)
15. Johnson, J.: CHAOS Rising, Standish Group. In: Proceedings of the 2nd National Conference on the IT System Quality (2005)
16. Panorama Consulting Group: 2008 ERP Report, Topline Results. Denver (2008)
17. Parthasarathy, M.A.: Practical Software Estimation: Function Point Methods for Insourced and Outsourced Projects. Addison Wesley Professional, Reading (2007)
18. Standish Group: CHAOS Summary 2009. West Yarmouth, Massachusetts (2009)

Chapter 18
The Role of Automation Tools Supporting the Maintenance of the System Environments

Wojciech Filipowski and Marcin Caban

Abstract. Tools to automate the maintenance of the system play an important role in managing a complex environment with high dynamics of change. Among such tools are: open tools - functioning under a public license, and closed solutions - which the source code is not published, and the usage is associated with the need to pay the fee for the rights of copyright holders. In this article there is the comparison of the two types of tools, together with identification of their use.

18.1 System Environments

Consider the case of a Corporation, which operates an integrated environment of information systems supporting the operational activities of the company by automating business processes and system integration of so called legacy systems performing specific functionality. Suppose that in this environment, a customer service system is running with a number of interfaces to communicate with other systems through the middle layer - called the integration layer. In this system, together with any edition of an integrated environment, there are changes - improvements, modifications, patches - implementing the business requirements. In order to implement such changes, each of the integrated system should be equipped with several environments that meet a number of precisely defined roles in the process of implementation and testing, to achieve best quality and reliability of a new version of the production environment at the time of starting a new business version of Corporate integrated environment. These Environments are at the responsibility of different actors conducting the project of the change implementation. Suppliers responsible for delivering new functionality need their development and testing environments,

Wojciech Filipowski · Marcin Caban
Silesian University of Technology, Faculty of Automatic Control,
Electronics and Computer Science, Akademicka 16, 44-100 Gliwice
e-mail: Wojciech.Filipowski@polsl.pl

E. Tkacz and A. Kapczynski (Eds.): Internet – Technical Development and Appl., AISC 64, pp. 157–164.
springerlink.com © Springer-Verlag Berlin Heidelberg 2009

particular units of the Corporation, carrying on different class of tests, the maintenance supplier needs his maintenance environment, and last but not least there is the production environment, where the operation processes are carried out and necessary data collected to support the operation of the company. Figure 18.1 specifies main classes of the change implementation support and maintenance environments of the system.

Short name of the environment	Full name of the environment	The role of the environment
DEV_TEST	Supplier's development environment	Software implementation and supplier's internal testing
SYS_TEST	Supplier's system test environment	Supplier's regression and / or system testing. Stand alone environment with simulated system communication via the interfaces
INT_TEST	Integration test environment (in the responsibility of the Organization)	Fully integrated environment for cross system business process testing
ACC_TEST	User acceptance test environment (in the responsibility of the Organization)	Fully integrated environment for user acceptance testing
PERF_TEST	Performance test environment (in the responsibility of the supplier or the Organization)	Performance testing environment. Can be integrated, depending on the adopted strategy and test scenarios
PROD_TEST	Production support test environment (in the responsibility of the Organization)	Fully integrated environment, for testing prior to the production version upgrade
BIZ_OPER	Production environment (in the responsibility of the Organization)t	Environment for performing business operations of the Organization

Fig. 18.1 The main classes of system environments

18.2 Selected Aspects of System Environment Maintenance

The problem of system environment maintenance, especially in large organizations having an integrated environment, is a task which depends on two very important processes: the process of maintaining the production environment and the process of implementing changes to the system. Since the maintenance process is a permanent one as we have to deal with it on a regular basis, even during the implementation process, these two interact continually. In the process of the production environment maintenance there are patches installed, which are a de facto functionality, performance or configuration modifications, so in most cases, changes in the application code. This situation causes the need to update the newly implemented application code to make sure that fixes to the production environment are included in the new version of the system. From the operational point of view, the maintenance of the system environment is to ensure its operation and availability of the appropriate version for users at the required level. This article addresses issues related to system environment maintenance, and the question of choosing a proper maintenance model depending on a class and purpose of certain system environment. In the Organization taking advantage of an integrated environment the process of implementing changes in this environment is relatively complex, and ensuring the correct version of the application code on each environment is also a complex matter. In other words, the case is to ensure an adequate flow of the application code between system environments. Figure 18.2. shows an example of the flow of code between the environments of customer service system in a Corporation using an integrated environment.

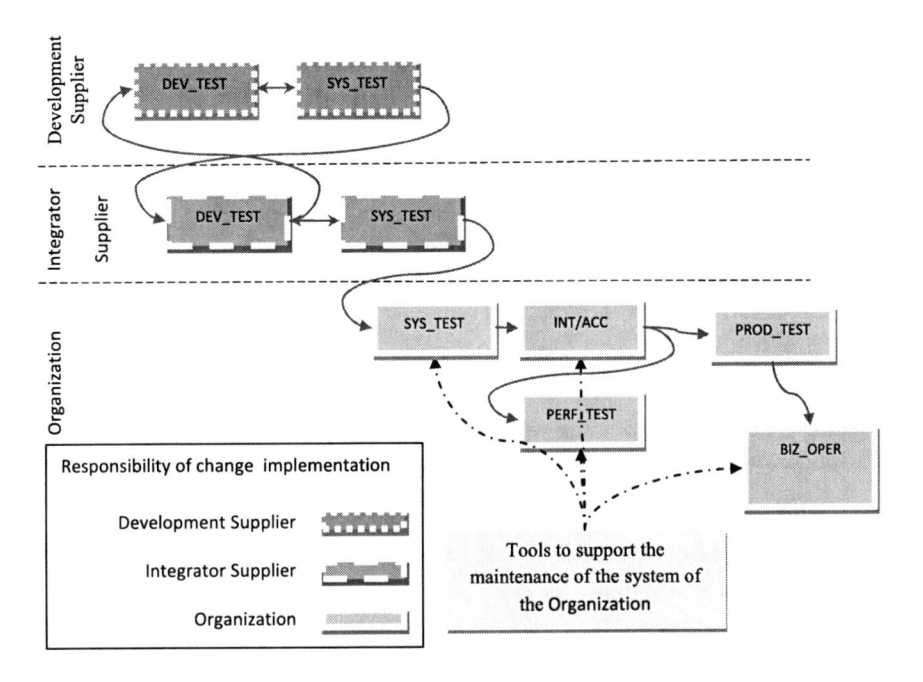

Fig. 18.2 Diagram of code flow between environments of the sample system

This example illustrates the flow of application code in the change implementation model of multiple suppliers with Development Supplier and Integrator Supplier. The further part of this article shows tools used by the Organization in the process of system environment maintenance.

18.3 Tools to Support System Environment Maintenance

To ensure proper handling of system environment versioning, certain tools are used to support the maintenance of these environments. Depending on the model of code management implemented in the Organization, those tools may be used in varying scope of their functionality. Overall, it is assumed that the scope of the functionality of tools for environmental change automation covers the following activities related to the management of system source code:

- Source code storage;
- Compilation;
- Build;
- Implementation,

but in real terms, depending on the model adopted in the Organization, the tools could comply with all or some of these tasks. Due to their construction and licensing tools can be divided into:

- tools that are of closed architecture, i.e. tools designed for a specific system and specific organization by the supplier. In this case, we are referring to tools designed in such a manner, that the supplier provides only functionality and does not provide the source code or architecture. In many cases, those tools are of simple and limited functionality closely matched to the requirements of the system operating in the Organization. Moreover, the configuration of these tools is often limited, with the result that in case of new requirements the Organization could provide to the supplier, the tools must be updated and a new version tools is to be provided. Although the unit cost of such tools is rather not a problem, but the unique nature of the tools causes, that the total cost associated with their use should be included in the budget of the Organization,
- open architecture tools, often licensed on an open source license, are yhe tools of publicly available architecture, or even source code. These tools are very flexible, provide a wide range of available functionality and the possibility of large-scale configuration. Due to their open nature, these tools can be relatively easy to integrate with each other creating a complementary set of tools for automation of system environment maintenance . The cost of the tools in this category is primarily dependent on licenses (in particular, free).

In the following part of the article, consideration will be carried out, related to supplier competence vs. Organization competence in the maintenance of system environments by using tools developed in a closed concept. Attention will be focused on selected elements of the process, such as the location of the source code repository, the functional scope of tools to assist the maintenance of the system environments and finally the role and tasks of the Organization and suppliers in the environment maintenance process. Figure 18.3. provides an overview of the key competencies of the supplier and Organization in the process of system environment maintenance by using tools developed in the closed concept.

If the system source code is located at the supplier's repository, and the Organization's repository is periodically replicated from the supplier's repository, the effect of the situation is basically the total dependence on one supplier for all areas of development and maintenance of the system. In addition, there should be attention paid to the phenomenon of hidden outsourcing, where according to the Organization

Area of competence of the supplier	Area of competence of the Organization
Implementation and maintenance of the source code repository	Administration of the System
Implementation and maintenance of developer tools and tools to manage the repository of code	Deploy packages with code provided by the supplier on system environments
Compilation and build of the source code for supplier's and Organization's environments	Responsibility in the context of a formal organization procedures without the necessary technical competence, which pertains to the supplier
Development and provisioning of tools for the Organization for code management and environment upgrade	
Update of the Organization environment upgrade tool	
Version management of supplier's an Organization's environments	
Responsibility for the correct update of the Organization environments	
The reference source code under control of the supplier	

Fig. 18.3 Comparison of competence of the supplier and Organization in the process of system environment maintenance using the closed concept support tool

procedures, all the responsibility for for the development and maintenance of the system belongs to the Organization internal units, and in fact almost all the powers are on the supplier's side, who is in charge of the efficiency and quality of accomplishment of these procedures. Therefore, introduction of modern tools seems to improve the efficiency of the process of system maintenance, each of which is well known in the solution market and their integration allows you to implement all the functionality of the process, that is: compilation, build and implementation. Moreover, thank to the fact that many suppliers know these tools, the organization opens up the possibility of diversification of the system area for IT solution providers, especially providers of system maintenance support. Figure 18.4 provides an overview of the key competencies of the supplier and Organization in the process of system environment maintenance with the open concept tools.

If the reference source code is stored in the Organization's repository from where the provider's repository is variable replicated, to implement the new system version (update), it requires that the supporting tools cover the following areas of functionality: compilation, build and implementation of the code to the environment (deploy). The following Figure 18.5 presents a summary of the automation tools to compare their main features.

Because of the wide scope of adoption of these tools, for their use Organization must dispose a qualified staff with very good knowledge in the environment maintenance processes and tools. When the tools are properly configured and implemented in the Organization the role of supplier is reduced to provide Organization source

Area of competence of the supplier	Area of competence of the Organization
System source code made available to the supplier by the Organization	Implementation and maintenance of the source code repository
Development tools for the supplier's purpose	Reference source code under the control of the Organization and made available to suppliers if necessary
Supporting Organization in the configuration layer of supporting tools only	Implementation and management of environment maintenance automation tools and tools integrated with the repository of code
Version management of supplier's environments only	Compilation and build of the source code for the Organization environments
	Deployment of packages prepared by Organization's own forces, using tools integrated with the Organization's code repository
	Formal and substantive responsibility in the context of the Organization procedures
	Keeping actual competencies in the area of source code management and system environment maintenance

Fig. 18.4 Comparison of competence of the supplier and Organization in the process of system environment maintenance using test automation tools developed in the open concept

code repository with code of best possible quality - merged and tested. This reduces the dependence of suppliers and ultimately improves the process of managing of multiple environments of the system.

18.4 Summary and Conclusions

The process of maintenance of the system environments belongs usually in the Organization to the group of supporting processes. This classification may lead to neglect the role of the tools and ignore the differences between them, regardless of the adopted model of system environment maintenance. Meanwhile, selecting the maintenance model with the most appropriate instruments can greatly help the Organization to manage complex environment as well as improve the position of the Organization in its relations with suppliers of IT services.

The Organization aware of its potential, is likely to significantly increase its competencies and negotiating position in relation with suppliers, by taking over part of the process of system environment maintenance from suppliers and implement low-cost tools developed in the open concept. All of that does not necessarily affect the relationship with suppliers, because it does not reduce the role of the supplier in

Feature	Feature coverage				
	Subversion	TortoiseSVN	Maven	Continuum	Archiva
Open Source license	x	x	x	x	x
Source code Repository	x				
History of changes	x				
Import data to the repository		x			
Working copy check out		x			
Commit of changes to the repository		x			
Creating and tagging of code branches		x			
Integration with bug tracking and development tools		x			
Build and implementation of the entire application			x		
Build and implementation of individual modules			x		
Version management			x		
Build and implementation report generation			x		
Automation of build and deployment				x	
Build history				x	
Unit test reporting				x	
Artifact repository					x

Fig. 18.5 Automation tools and their features

a key area of IT solution provisioning, but even release the supplier from the responsibility for the supporting processes in the implementation of changes to the system. At the same time, deciding to move to open tools and taking the responsibility for the reference source code the Organization creates the competition space for suppliers. Thus, the appropriate strategy in the area of maintenance of the system environments is important from the business point of view, although the assistant - from the technical point of view, as for the Organization operations. The concept of open tools should be seen as a trigger of development for the Organization in the following spaces:

- Technical - since it provides new tools to effectively manage an increasingly complex system environments;
- Business - because it gives an opportunity for more effective management of investment funds, through the diversification of IT service providers.

References

1. Ferguson, J.: Smart. Java Power Tools. O'Reilly Media Inc., Sebastopol (2008)
2. Li, S.: Introduction to Apache Maven 2. DeveloperWorks (2006)
3. Collins-Sussman, B., Fitzpatrick, B.W., Michael Pilato, C.: Version Control with Subversion, O'Reilly Media Inc., Sebastopol (2004)
4. http://subversion.tigris.org
5. http://tortoisesvn.tigris.org
6. http://maven.apache.org
7. http://continuum.apache.org
8. http://archiva.apache.org
9. http://sourceforge.net

Chapter 19
Why IT Strategy Does Not Suffice?

Andrzej Sobczak

Abstract. As it results from studies carried out by the author, more and more government units in Poland has developed IT strategies. However, it appears that commonly these materials have purely declarative character and are not reflected by real actions — especially by information projects being realized (in compliance with a rule that the urgent things prevail over the important things). Considerations on how to perform the operationalization of IT strategy with use of a concept of Enterprise Architecture were undertaken in the article.

19.1 Introduction

Nowadays, no one ask the question **what for** to perform informatization of government. However, more frequently there appear questions: how to do it in an effective manner in order to dispose of public funding in a rational manner and, simultaneously, to ensure that IT solutions being created be matched to needs of both government units and its customers, that is citizens, enterprises and non-government units, as far as it is possible. This is all the more important since within a few coming years Poland will receive from EU significant financial contribution for implementing a widely understood concept of information society (including e-government).[1]

A starting point for answering these questions is a diagnosis of a current state within issues concerning: an informatics role in government units, methods of organizing IT personnel in these units as well as approaches applied for developing new information systems necessary for government. For this purpose, the author carried out poll studies during II–IV quarter of 2007, and then profound interviews with

Andrzej Sobczak
The Warsaw School of Economics, Department of Business Informatics
e-mail: sobczak@sgh.waw.pl

[1] For example, there is spent about 2.3 billion EUR on the priority "Information society establishment and development" within Innovative Economy Operational Programme 2007–2013.

E. Tkacz and A. Kapczynski (Eds.): Internet – Technical Development and Appl., AISC 64, pp. 165–174.
springerlink.com © Springer-Verlag Berlin Heidelberg 2009

representatives of selected government units (IT department managers and directors). Studies regarded all ministries, voivodeship offices, marshal offices as well as
selected local government units. As a criterion of selection of local government units
it was decided to use results of contests for the best local government units organized
since 2003 by the Association of Polish Counties. Counties, cities with county rights
and communes participating in the contests of the Association of Polish Counties
undergo a very detailed evaluation through a wide set of criteria. By meeting each
condition they receive a specified amount of points (i.e. possessing the ISO certificate 9001:2000 — 350 points, the certificate of participation in "Przejrzysta Polska"
— 200 points). It was decided that a poll questionnaire will be directed to 100 best
units in a contest (presuming that there are more than 100 units because one position in a contest may be occupied by more than one unit). In general, there were sent
about 330 questionnaires responded by 31,6 % of units.

In the article there were also presented results of pilot studies carried out in I
quarter of 2009 by the Department of Business Informatics in the Warsaw School of
Economics[2] and the Sybase Polska company. These studies had character of structured phone interviews with IT departments directors or managers. In order to ensure
the unified interpretation of questions by respondents each question was enriched
by necessary definitions and an explanatory comment. The study itself was divided
into two parts. The first one — general — focused on discovering a degree of describing business processes, enterprise data model and information systems among
the analysed organizations. Furthermore, respondents answered questions concerning problems occurring when business and IT have a close contact with each other
and related to an approach to implementing IT solutions. The second part of the
study concerned issues directly related to Enterprise Architecture — a degree of
knowledge about notion itself and applying concrete approaches to building architectures. In general, there took part 96 subjects representing companies from the
Rzeczpospolita newspaper list gathering 500 largest enterprises as well as representatives of ministries and central units.[3]

19.2 Attempt to Evaluate the Selected Results of Poll Studies

For purpose of this article, results of poll studies concerning issues related to IT
strategy in public organizations will undergo a detailed analysis.

Figure 19.1 depicts respondent answers for a question: "Does your organization
have assumed IT strategy" depending on the respondent size. It shows that the larger
an organization the more often it has developed IT strategy — and in case of very
large organizations it is (or will become in the nearest future) even a standard.

---

[2] The author was responsible for technical preparation of the studies on behalf of the Department
of Business Informatics in the Warsaw School of Economics and carried out an analysis and
interpretation of the received results.

[3] For purpose of this article results concerning solely government units will be presented.

Simultaneously, while analysing IT strategies of government units (especially with regard to local government units) which are available to the public within public information bulletins, it can be observed that frequently they are largely similar to each other. On the one hand, one can state that it is hard (or even impossible) to "reinvent a wheel" each time — and all the more since public units operate within the same legal regulations and have the same public tasks to realize. Simultaneously,

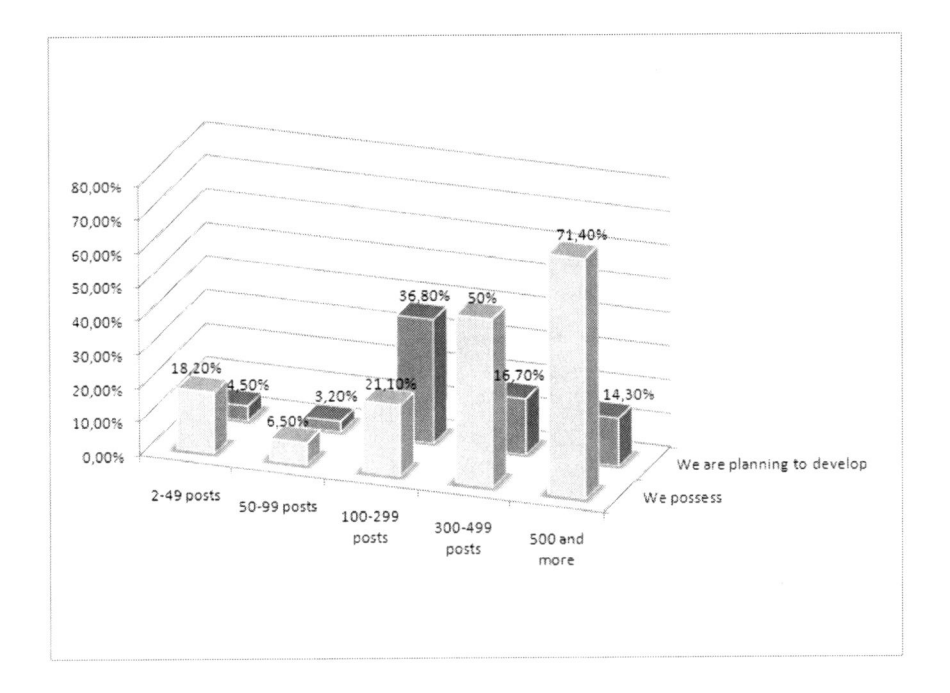

Fig. 19.1 Organization's disposition of IT strategies (an analysis including the size of an organization)

Source: Own work on the basis of results of the poll studies from 2007

one can feel some insufficiency within three areas. Firstly, strategy written in a generic manner (very often in its subsequent occurrences there change solely contact and localization data of an unit as well as results of the made IT resource inventory) is not real support for information projects planned for realization by an office — it becomes de facto a typical "thing outside the mainstream". Secondly, there is neglected a possibility of identifying information solutions which will be aligned to needs of an unit and its customers, that is enterprises (current and potential) and citizens. Finally, there are no chances of implementing systems having an innovative character enabling one office achieve a competition advantage (in contrary to the common opinion that public organizations frequently appear to function as monopolists within production of goods and provision of services while, for example, between local government units most of all there is a rivalry for investors as well as for positive perception of a city/county/voivodeship in a scope of a whole country). In this last case, one should tell about a necessity of developing an innovation strategy with use of IT or about a society development strategy based on knowledge (most of all such tasks should attract interest of large cities).

Above considerations allow to form a conclusion that within Polish government units there emerges a gap between informatization strategies being formed and information projects and programmes (considered as a set of projects) being realized by offices — see Figure 19.2.

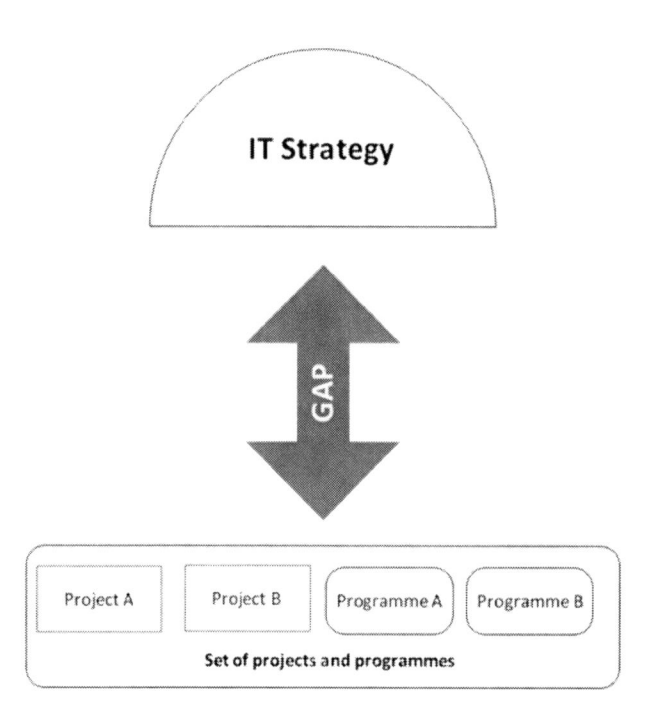

Fig. 19.2 A gap between an IT strategy and projects and programmes being realized in offices

Source: Own work

In practice it takes place in such a manner that a strategy is formed, but it has such a general character that it is hard to address during a realization of concrete IT systems. It implies that information systems being realized within these projects and programmes does not support strategic organization goals. This conclusion has been confirmed empirically by the author within both mentioned studies.

During realization of studies in 2007 respondents were asked to determine from their point of view the importance of each feature during creation of new information systems using a scale $1-10^4$. The mentioned features included: system implementation speed, wide functionality of a system, compliance with legal regulations in force, easy integration with currently exploited systems, use of high technology within a system, friendly and easy handling, agility and scalability of a system, small hardware requirements, data processing speed, support for the assumed strategy, alignment of an information system to an institution management system, provision of data security, compliance with standards and a system purchase price. According to respondents it appeared that "support for the assumed strategies" is the least important factor — see Figure 19.3.

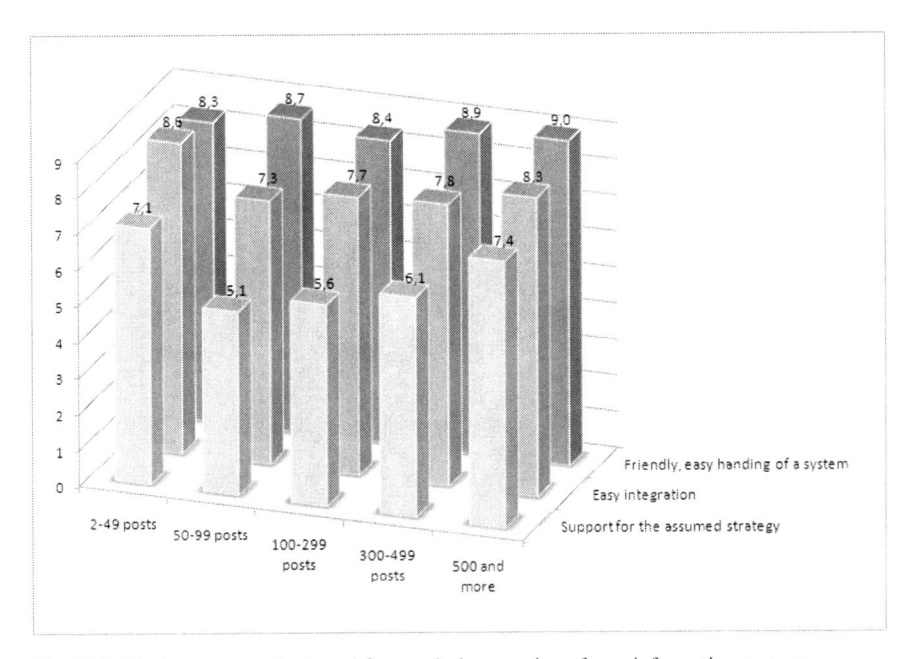

Fig. 19.3 The importance of selected factors during creation of new information systems

Source: Own work on the basis of results of the poll studies from 2007

Much more important issues regarded friendly handling or easy integration with currently existing information solutions. It is a confirmation of a rule that the urgent things prevail over the important things. However, such an approach leads to creation of suboptimal solutions (locally optimal), that is a situation when information systems being created realize needs of the separated part of an organization (i.e. department/section) but its operation does not support realization of strategic IT goals. It can result in emergence of difficulty in long-term maintenance and development of systems in a scope of a whole unit.

---

[4] 1 — stood for the least important feature, 10 — stood for the most important feature.

Results of the pilot studies in 2009 also confirmed the gap diagnosed beforehand. Respondents were asked to evaluate a level of alignment of IT systems to strategic goals in force in scale 0–5[5]. It appeared that 42.8% of respondents indicated that there is no such alignment or that this alignment is relatively poor (herein included indications of respondents which had value of 0, 1 or 2). It is also symptomatic that no respondent believes that information systems in his/her office are fully aligned to the assumed strategy (see Fig. 19.4).

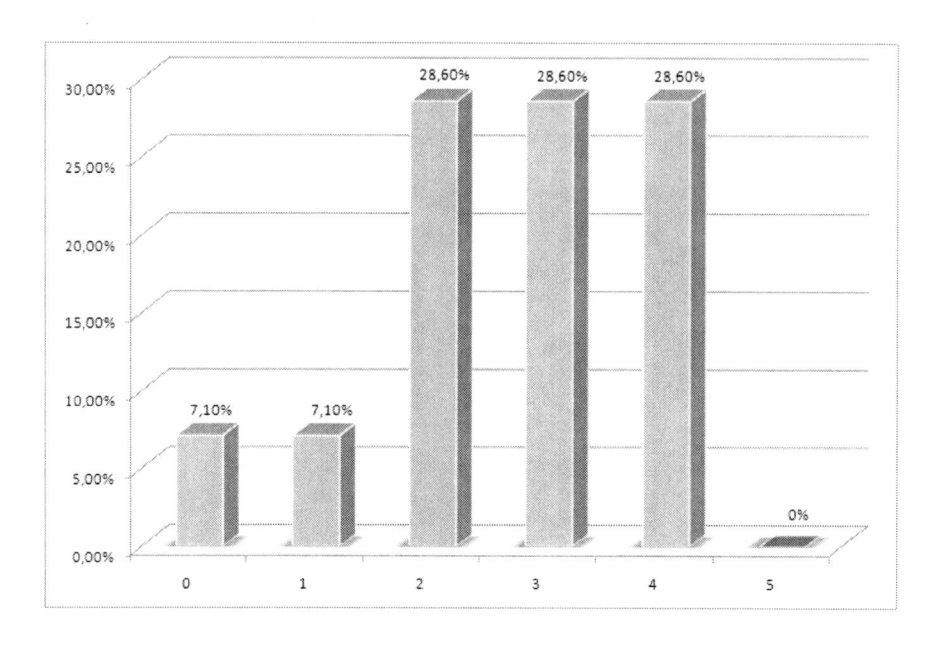

Fig. 19.4 A level of alignment of IT systems to strategic goals in force in government units

Source: Own work on the basis of results of the poll studies from 2009

19.3 Enterprise Architecture as a Basis for IT Strategy Realization

A tool which allows to decrease or even eliminate the gap diagnosed in the previous point is Enterprise Architecture [3] — see Figure 19.5. For purpose of the US government there was defined Enterprise Architecture as a strategic organization information resource within which its mission, information and technical resources necessary for realization of this mission as well as a transition process intended to implement new technical solutions in response to strategic changes in organization [1] are defined.

---

[5] 0 — stood for lack of alignment, 5 — stood for full alignment.

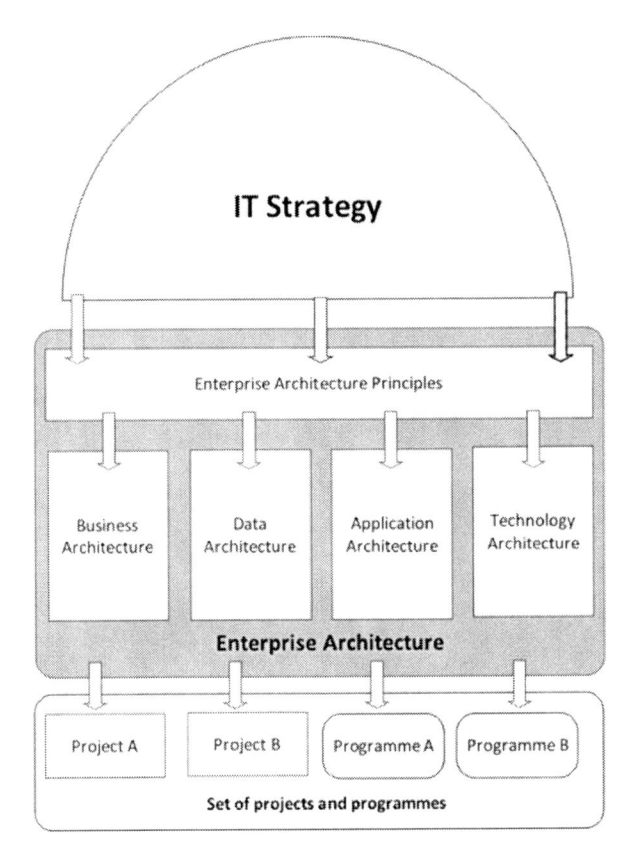

Fig. 19.5 Minimization of a gap between an IT strategy and projects and programmes being realized in offices with use of Enterprise Architecture

Source: Own work

Enterprise Architecture has a form of a set of models which comprises of reference architecture — in the literature named as "as-is", target architecture — in the literature named as "to-be" and a transition plan being a strategy of an organization change within a transformation of its reference architecture to target architecture [4, 7]. The Open Group indicate in its elaboration that Enterprise Architecture comprises of following elements [8]:

- Enterprise Architecture Principles, which is a set of durable principles based on an organization strategy that are a representation of holistic needs of an organization within creation of information solutions;
- Business Architecture, which documents a business strategy and ways of managing an organization, its organizational strategy as well as main business processes and relations between these elements;

- Data Architecture, which describes main types and data sources necessary for operation of an organization;
- Applications Architecture, which describes software systems, its location, mutual co-operation and relations between these systems and main business processes of an organization;
- Technology Architecture, which describes technical infrastructure that is a basis for operation of key software systems (it includes operational systems, database management systems, application servers, hardware and communication infrastructure).

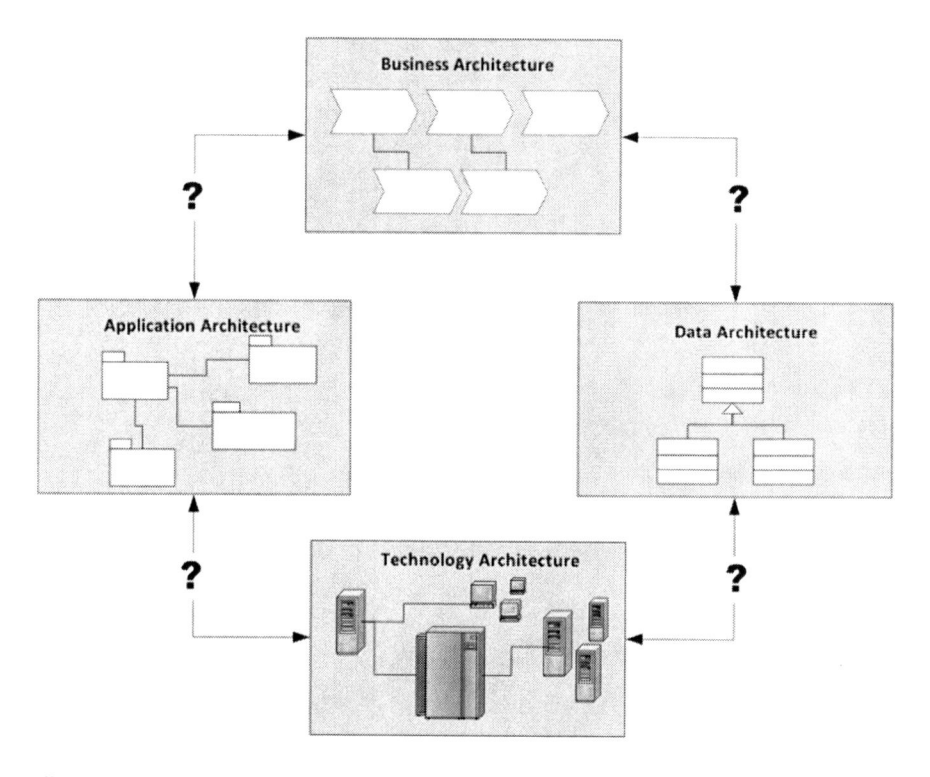

Fig. 19.6 Relations between components of the Enterprise Architecture

Source: Own work

A key factor deciding on the power of Enterprise Architecture are relations between components of Enterprise Architecture — see Figure 19.6. Owing to this, there is a possibility of coordinating several aspects of organization operation in a holistic manner. It ensures also a possibility of efficient alignment of IT systems to an organization strategy and improvement of allocation of resources spent on development of information solutions.

In addition, A. DiMaio believes that Enterprise Architecture is a tool supporting an organization transformation process being realized on the basis of the assumed strategy. The transformation itself is defined as changes in processes and organization frequently supported or expanded by information solutions leading to realization of strategic unit goals. He gives examples of governments that have applied this approach for some years: Canadian, German, English and US. [2, 5].

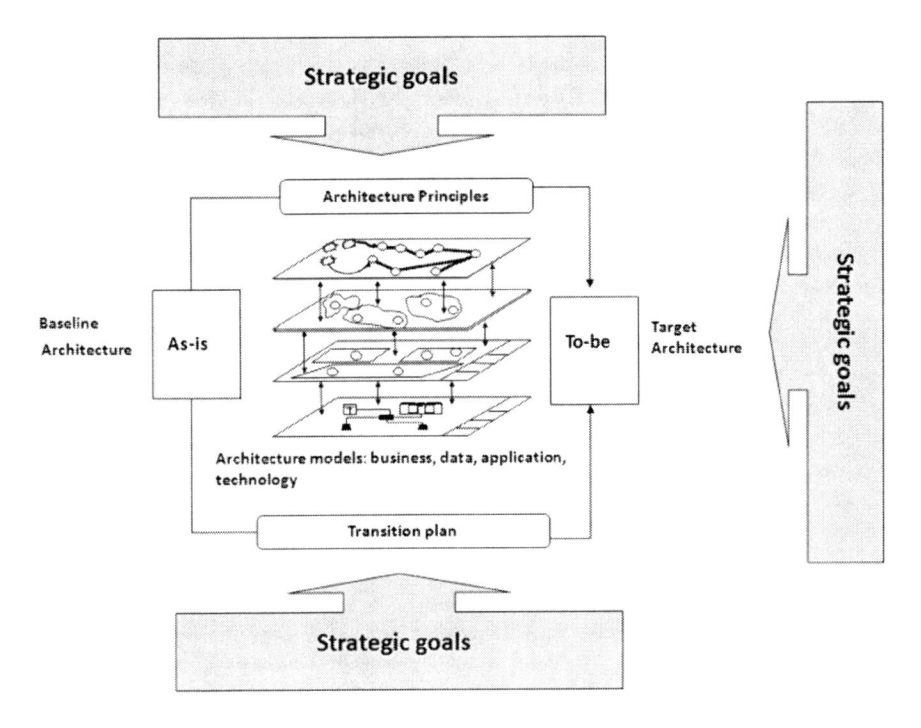

Fig. 19.7 Strategy operationalization with use of Enterprise Architecture

Source: Own work

Figure 19.7 depicts a schematic view on strategy operationalization with use of Enterprise Architecture. A starting point is determining reference architecture, that is settlement what an unit disposes of in a given period of time (within all four architectural domains: business, data, software and technical infrastructure). Then, on the basis of a strategic goals analysis target architecture (also within four architectural domains) is defined. It is a basis for creating a transition plan (realization projects and programmes related to each other). Its realization includes the architectural principles assumed by an organization.

From such a perspective, Enterprise Architecture has a role of a filter allowing to evaluate real resources necessary for realization of each strategic goal. From the experience of the author it results that a strategy in public units is frequently only a set of not realized expectations within existing resources.

19.4 Summary

An issue of Enterprise Architecture is still relatively little popular in Poland - on both research and application layer [4]. A purpose of the considerations discussed in the article was an attempt to draft applying this concept as a tool supporting realization of IT strategy. Owing to Enterprise Architecture it is possible to prevent creation suboptimal solutions (locally optimal), that is situation when information systems being created realize needs of the separated part of an organization (i.e. department/sector) but its operation does not support realization of strategic IT goals. It has the significant importance in case of government units because this concept enables rational management of public funding spent on informatization [6].

References

1. A Practical Guide to Federal Enterprise Architecture. Chief Information Officer Council, version 1.0 (February 2001)
2. DiMaio, A.: Government Transformation. In: Enterprise Architecture and Portfolio Management: Which Comes First?, Barcelona, Spain, Gartner Symposium IT Expo., May 21-24 (2006)
3. Ross, J., Weill, P., Robertson, D.: Enterprise Architecture as Strategy: Creating a Foundation for Business Execution. Harvard Business School Press, Boston (2006)
4. Sobczak, A.: Architektura korporacyjna — pojecie, geneza, korzysci, w: Wprowadzenie do architektury korporacyjnej, red. B. Szafranski, A. Sobczak. Wydawnictwo Wojskowej Akademii Technicznej, Warszawa, s.21–s.36 (2009)
5. Sobczak, A.: Transformacja polskich organizacji publicznych z zastosowaniem koncepcji architektury korporacyjnej, w: Wprowadzenie do architektury korporacyjnej, red. B. Szafranski, A. Sobczak. Wydawnictwo Wojskowej Akademii Technicznej, Warszawa (2009)
6. Sobczak, A.: Wybrane aspekty zastosowania architektury korporacyjnej w organizacjach publicznych, Centrum Promocji Informatyki. Elektroniczna Administracja — dwumiesiecznik o nowoczesnej administracji publicznej 1(14), s.29–s.42 (2008)
7. Sobczak, A.: Zastosowanie modeli i metamodeli w architekturze korporacyjnej, w: Wprowadzenie do architektury korporacyjnej, red. B. Szafranski, A. Sobczak. Wydawnictwo Wojskowej Akademii Technicznej, Warszawa, s.37–s.56 (2009)
8. The Open Group, The Open Group Architecture Framework ver. 9 (February 2009)

Chapter 20
The Process Approach to the Projects of the Information Systems

Vladimir Krajcik

Abstract. This article presents issues related to application of process approach to the information system projects.

20.1 The Processes in the Projects of the Information Systems

From the perspective of a general determination of a concept we can apply several definitions. The most general ones result from the interpretation of the process as an organized whole, which receives inputs and transforms them into outputs. V. Repa [13] defines then the process as a sum of activities transforming a sum of inputs into a sum of outputs for other people or other processes, along with using people and instruments accordingly. According to V. Svata [14] from the Department of Information Technologies of the University of Economics, Prague the process represents a set of follow-up activities, which create the desired output from the defined inputs, relate to each other resources and show some measurable characteristics. In the stated definition the fundamental key element of the process is the activity. From the semantic relativity principle [1] (which comes from the fact that the primary type of the hierarchical abstraction in the process structure is aggregation), we can deduce a statement according to which each activity can be described as the process. The process approach to the activity and its description is then dependant on the need of the understandability of the model, the applied instrument, the invention and style of the describer of the model, as well as the model size limitation.

The mutual interaction takes place among the processes. The state starts and ends with an event, which relates to the performance of the subsequent process. It is possible to introduce a relation of precedence and create the structural relations of the process arrangements into the particular process groups, characterized in the

Vladimir Krajcik
Business School Ostrava
e-mail: Vladimir.krajcik@vsp.cz

E. Tkacz and A. Kapczynski (Eds.): Internet – Technical Development and Appl., AISC 64, pp. 175–184.
springerlink.com

graphic form by linear or branch connecting lines. **Such defined descriptions of the processes necessarily imply the specification and definition of the project process as the transformation of project inputs into project outputs by means of activities with a precisely delimited relation (relations) of succession.**

By way of this definition I am heading for the understanding of a very important method and practices of process management - the straightening and integrating of partial processes into a whole both on the horizontal and also vertical level.

The horizontal level is easy to comprehend and graphically demonstrable by means of the relation of succession. Its formulation enables a superior monitoring of the progress of the project performance, time, contents and organization, i.e. the supervision and evaluation of project working teams within the project activities. The horizontal direction contributes to the specification of the interface among the processes and makes the pace of the implementation of processes fast. It is an expression of the natural sequence of particular activities in the global process of project activities.

The vertical level is more abstract, yet more significant to the processes of strategic project management, its objectives, plan creations, control processes and project change management, together with setting project priorities as a consequence of the implementation of decision making processes. The vertical direction contributes to the faster procedure of project work in the need of strategic decisions (these decision making processes are incorporated and integrated in the process groups) along with receiving and accepting project changes and it is a significant element for the definition of the expression of function, requirements, and goals of the project.

Other approaches to the delimitation of the concept of the process are emphasized by some of its more general specifics. For instance, a PMBOK methodology (A Guide to the Project Management Body Of Knowledge) [9] defines the process in relation to knowledge used by it. Each field of knowledge is then described in detail by characteristics and application. The structure of the process is as follows:

- Objectives - include the specification of the process and its sense.
- Inputs - determine materials and requirements, which are necessary for fulfillment.
- Outputs - include the same specifications as inputs.
- Instruments and techniques - refer not only to interesting procedures and modern management methods but also to qualification requirements for the participants in the given process.

In the concept of the process in relation to the project I highlight the relation of the process and change. The process [8] is an abstract term designating changes of qualities of objects and subjects. The outcome of processes is the changed objects and observed phenomena. The process is an action or event, which takes place with objects at a specific time - the process is then dynamic.

When defining the process other authors emphasize the target direction of the process and attach essential importance to this concept. According to [3] the process is a discrete system, which is initiated by the event, which sets activities in run so that the system would move towards the target state. The delimitation of the target state is in this concept rather problematic (here it is understood as stochastic), and curbs permanent changes in the system.

The general need of the change (on various levels) generates processes. The project concept of processes reflects the need of the continuous improvement of partial processes such as instruments for providing project acquisitions [6].

20.2 The Classification of Processes

When classifying processes in the context of the solution of the IS projects it is necessary to define and specify the basic categories of processes in the general project understanding. The basic classification defined in this way will enable me to understand and classify processes in the context of the solution of the IS projects.

From the perspective of process outputs, i.e. the creation of the added value of the product we can distinguish:

1. Key processes - the added value arises in these processes, which serves to satisfy the need of the external customer - the user of the product. The main processes do not usually exceed the boundaries of the particular areas, they specify rather one area.
2. Supporting processes - the added value arises in these processes, which serves to satisfy the need of the internal customer. It is a strategic or critical product (service), which cannot be provided externally. The threat of the mission of the enterprise could take place with the external provision.
3. Side processes - support key processes in the enterprise by its character and are connected with activities of their provision. Their contribution lies in the fact that they enable main processes to be conducted with an optimal performance. These processes can be implemented via outsourcing (following the economic evaluation of their efficiency and cost aspects).

The above listed process groups in the IS projects implement product outputs in their final stage with the added value for the customer - i.e. an information product used by a consumer - a user for the purpose of satisfying the specific information needs. However, non-production processes operate above those processes. Non-production processes do not have that feature, nevertheless they are in their essence indispensable for their own application and performance of production processes. They have a strategic significance both in the company process level, and also in project understanding of the IS projects. In enterprises, they provide for the long term prosperity of the company, in projects (especially IS projects) they maintain the long term sustainability of project outputs (information systems), the accord of the project strategy with the institutional strategy and strategic documents (societal and social ones etc.). I rank the process of management among the essential non-production processes, (in the project context it is project management), along with monitoring (in the project context it is a process of project monitoring), decision making processes (in the project context it is a process of project decision making), and control processes (in the project context it is a process of project control).

Non-production processes in the IS projects always take place by means of purposeful activities. The purposeful activity is always carried out in a relatively closed system (project system), which has to have a state given by its structure. What follows from that is the process of management in the IS projects has to contain both the creation, the maintenance of the system structure and also the management of internal dynamic processes. Therefore we can distinguish a process of management in a broader sense by applying the principle of abstraction, which includes a process of strategic management, management of particular partial units (e.g. processes of financial management, management of logistic processes, processes of management of human resources project capacities), change management, processes of regulation (i.e. processes of monitoring and control), and a process of management in the strict sense (the process of one's own management in a system). The set of processes of a broader determination of management fills to a substantial degree the set of non-production processes.

20.3 General Principles and Characteristics of the Process Approach to the Management of Projects of Information Systems

If we recognize the process elements of the IS projects (including the respective IS methodologies) and in particular the process concept for managing projects of information systems as a way of providing efficient management of the IS development, then from the system point of view it is essential to define general principles of this viewpoint and process approach to the IS projects - as well as to define principles and rules of the process approach. To put it differently, we are delimiting basic rules of the approach for setting the project approach and contents of the lifecycle of the IS, a degree of generality with respect to the degree of specificity and alternativeness of the given specific solution (methods, techniques, tools) in the given specific project stage.

At the same time, it is necessary to define the process approach to managing the IS projects in relation to the requirements of the customers for the continuous outputs of these information projects. The competitive environment is a strong factor, which constitutes positive pressure on the continuous evaluation and improvement of all processly managed projects. To be more specific, in the IS projects that means an emphasis on the repetitive and constant method from the description of processes to their measurement of efficiency, the effectiveness of operation and the use to a logical conclusion in the form of the proposal and implementation (continuously) of the new process groups in the project. This process development and process progress can be defined according to V. Repa [13] as " a natural process approach".

The following stated principles have become the starting and fundamental principles (axioms) in forming the principles of the process approach to the project solution of the IS projects, and the formulation of the elaboration of the methodology of the process approach to the management of the IS projects. We are dealing here with the following rules.

20.4 The Principle of the Process Strategy

The principle of strategy and strategic concept is the fundamental principle of the methods of the proposal and management of the information system. The process strategy shows that one's own project solution and project production activities are not the only process in the project; however there are also process groups, clusters, sets and perspectives related to them, which can be described by the specific project process model and moreover, they can be divided into the main and side process groups. These process chains influence and complete each other often take place pararelly, outputs of some them enter as inputs into others, and the revision of processes and change management affect other entire branches and are subject to control mechanisms.

The strategy of applying process procedures and features of their tools and techniques come from the idea that the information system is a model of a real system (real world) and its basic sense is the creation of outputs for the final use by the customer. The process strategy integrates necessarily in itself an element of purposefulness - processes transform inputs into outputs for the purposeful use. The purpose or requirement can be viewed as a basic goal, defined by a broader project (strategic) context, in which the development of the information system is carried out. It implies that the project and system environment, yet inexplicitly, represents a reason or requirement for the given information output to exist, as it requires or needs it. This existing, unobvious or potential requirement is formed by the general process strategist (in the project concept of the project managing board) so that it would comply with the fundamental objective of all processes in the project, i.e. provide for superior project output - IS (and its sustainability on the market).

The mission of the IS project is significant from the perspective of the principle of the process strategy and formularization. Unlike the purpose of the IS project, the strategic project manager decides about the contents of the project mission and indirectly respective interested people (stakeholders) with their respective competency. The mission of the IS project responds directly to the purpose, expresses positive attitude to it, as well as it is to be fulfilled. A well processly formulated mission of the IS project accounts for the justifiability of the IS project existence and its outputs, it presents why and for whom it is here and in what way it can be useful for them.

20.5 The Principle of the Consequent Separation of Production and Non-production Processes in the IS Projects

I identify production processes in projects of the IS development with the processes, which carry out outputs with the added value for the customer in their final stage - i.e. an information product, which is utilized by the final consumer - a user for the purpose of satisfying specific information needs. On the other hand, non-production processes do not have that feature; however, they are crucial in essence for their own application and performance of production processes. In the IS projects (or in the process understanding of those processes) non-production processes are carried out throughout the entire period of the project duration, i.e. we can expect their launch anytime according to the preliminary requirements for the specific case of use of the given process and the schedule of process management. Production processes are more " stable" in this respect. Their start is given by the described process branch; their outputs determine the creation of further production processes.

The principle of the consequent separation of production and non-production processes in the IS projects and its application in the process model of the IS development by way of projects makes the first analytical classification of processes in projects possible. In the ideal circumstances, we can generally assume that by means of the consequent separation of partial processes into these two essential process groups we will obtain two disjoint process groups (breakdown classes of the basic set of project processes). This basic feature is maintained by another consequential partial incremental improvement of processes or partial separation into other project groups.

The fundamental objective of applying this principle is then the separation of performance towards a process object - the product (e.g. user documentation, as well as websites of the outputs of the IS project) from information for strategy and management of performance towards a product of the IS project (e.g. information stored at the website of one's own project management containing a list of key activities, time schedule etc.).

By observing the above mentioned principle, we will achieve some other features of the given process group. The owners of all processes in the IS project will be classified by the process approach into the owners of production and non-production processes. This is a prerequisite for the further research of these two disjoint classes.

20.6 The Principle of Delegation, Monitoring and Control

The basic sense of applying the principle of delegation and control as the process approach to the IS project management lies in the fulfillment of features and abilities of the product to be created by the project. These features refer to the product relations (of the particular IS) towards the environment. Information flows and elements of cooperation (interaction processes) are the connection of the objective of the IS project and its output. It practically means a diversion from the single objective - the maximal production efficiency of the project (i.e. an attempt to create a maximum number of information outputs as cheaply as possible) to the efficiency oriented on the product (project output of the IS project) and the customer. The fulfillment of the fundamental objective of the project in this concept is the consistent application of the process approach of delegation. What is always meant by delegation is the special-purpose handover of power and responsibility to the owner of the process, who can himself comprehend thoroughly and get to know the momentary need and if necessary apply the inevitable implementation processes leading to the fulfillment of the above mentioned mission of the process (or process groups). The structure of the process teams is arranged in such a way, so that it would be possible to delegate and adapt the team to the requirements laid on them. The processes of delegation are often accompanied by decision making processes, which bring along an element of responsibility for the acceptance and carrying out of impacts of the respective decisions in the follow-up efficient and implementary processes and launch other change management. For instance, if the period of the project duration is limited by the specific deadline of the handover of the information output, then in the process of risk management it is often necessary to accept a higher degree of risk in the processes of the scope of the project, project costs and testing processes with the aim of observing this determined project parameter.

The principle of monitoring and control in the process concept is based on the continuous knowledge of the course of the particular processes with the main emphasis on those processes, where the change is the most probable and necessary. It means constant attention to non-standard processes and project activities and an attempt to maintain outputs of those processes in relation to the defined strategy and project objectives. The need of the application of the principle of monitoring and control is the greater, the profounder the application of the principle of delegation is. Partial control processes are necessary to be planned in the respective cycles - it affects also processes of project planning and processes of the integrated control of changes. The significance of the application of this principle is high also in multiproject management and multiproject environment. The multiproject environment means that different projects in an organization take place pararelly and their solutions are often bound by the common provision of side and supporting processes. In addition to that, those projects are in different life cycles. Not all project managers and project workers are sufficiently equipped with knowledge and completely apply principles of process project management. In spite of it, what follows from the description of the implementation of multiproject environment is the necessity of process delegation. When applying it, it is necessary to observe the principle of the process approach of monitoring and control, whose indispensable condition is:

- The separation of the specified processes supporting the common process infrastructure,
- The use of the particular instruments and methodologies for project management,
- The launch of the process of setting priorities with particular parallel processes in the solved projects,

- The monitoring of interactions of particular processes within one project and in the multiproject context,
- The control of project limitations and risk procedures (impacts),
- The standardization in the processes of communication and information flows.

20.7 The Principle of the Repeated Procedure in the Management of the IS Project

The process of the gradual development of the product of the IS project is embedded in the project time axis. From the point of view of the project, the implementation of the product takes place in the particular project stages, phases and steps. The project procedure determined in this way is not linear, yet it brings problems and changes, which have internal and external project causes. The delineation of the process approach as a way of solving these situations requires the acceptance of the change and procedure variations too. The repeated procedure (or the introduction of the process of the repeated procedure) in the management of the development of the IS assists in identifying risks in each stage of the lifecycle of the project, and reduces considerably costs of their removal. For the purposes of management, the lifecycle of the project is divided into certain phases, and each phase is composed of a series of small parts of repetitions (iterations). The particular iterations include four fundamental activities: the collection of requirements, the proposal, the implementation and evaluation.

In the particular process stages, the requirements for the IS are analyzed, as well as refined merely to such a degree, so that it would be possible to propose a system for the launch of another stage based on the carried analysis. In this way it is possible to prevent from the useless details of the proposal at the beginning of the development and reduce the risk of consequent changes of requirements. Requirements are just put in detail. In the case of the change of the superior requirement, or by adding more requirements it is always necessary to return to the preceding stage, where the requirements appeared for the first time as the change. After that, the resulting changes from it can be projected into all subsequent stages.

20.8 The Principle of Team Work, the Principle of Evaluation and Motivation of Workers in the Project

Although the essential attribute of the process is its unique owner, the process activities and project actions are not carried out by individuals without the mutual interaction. In the process understanding partial processes create specific outputs, which enter into other processes and influence significantly their implementation. The mutual influencing is so strong that the project output and its parameters do not depend just on the simple total of individual performance but rather on the consistent application of the principle of team work. It is necessary for the proper performance of project management to accept and enforce the interests and the purpose of the process strategy, influence partial project teams and create the suitable process and organizational structure. However, the process approach does not neglect the emphasis on the autonomy of project teams with sufficient powers. Yet, the motivation of these teams for the project performance has to be in accord with the added value of the use of the output for the customer and user of the output of the given process. It is necessary to approach motivation complexly and unwind it from the fulfillment of the final product. This finding of facts can be supported by the process viewpoint of the entire solved project. The fundamental process (i.e. the whole project) is concluded following the fulfillment of outputs of all partial processes, which are carried out in the project.

20.9 The Principle of the Maximization of the Application of Knowledge Processes

The stated principle proceeds from the assumption that the IS projects and their output i.e. IS/IT products are the proof of a general trend of the transition of industrial economy to knowledge economy. The added knowledge value of IS outputs is so high that it is necessary to maximize knowledge processes in respective types of projects and their specific execution. Projects based on this principle enhance the total project intelligence (in the human factor, organization structure, managing units, control mechanisms . . .) and create prerequisites for processes of permanent learning (the model of the learning project.) The application of the knowledge principle simultaneously means the removal of the information and knowledge obstacles for their diffusion in the project - the opening of channels of " the flow of knowledge". This flow takes place both in the horizontal direction (across individual processes) and also in the vertical direction (across project structures), borders of project activities are getting loose. In this environment of the flow of knowledge, the owners of the process maximize the effect and efficiency of the controlled process in the project; moreover they maximize also the value of the product by this principle - the output of the IS project.

20.10 Conclusions

The process project approach emphasizes the application of principles of project management, knowledge technology, techniques, instruments and their application. It incites the need for solving non-standard processes in project management. These processes require another level of application of project managerial activities, the supervision and control over their implementation. Without the thorough knowledge of process approaches and the mastering of instruments for the support of general project management, project communication, computer modeling instruments, knowledge managment, it is impossible to provide processly for the above mentioned non-standard solution.

References

1. Chlapek, D., Repa, V., Stanovska, I.: The development of information systems. In: Oeconomica, Prague (2005) ISBN 80-245-0977-z
2. Chlapek, D., Choholaty, D.: Project management of IS/IT. In: Oeconomica, Prague (2004) ISBN 80-245-8O8O-7
3. Checkland, P., Scholes, J.: Soft Systems Methology in Action. John Wiley, Chichester (1990)
4. Hwang, C.L., Lin, Y.J.: Group decision making under multiple criteria. Springer, Berlin (1987)
5. Krajcik, V.: Information Center for Entrepreneurs - Processes and Project Management. In: 4th International Symposium International Business Administrations. Silesian University i Opava, Karvina, pp. 382–391 (2006) ISBN 80-7248-353-6
6. Krajcik, V., Ministr, J.: The analysis of processes and project management of the public IS Information points for entrepreneurs. In: System integration 4/2005, Czech society for the system integration in Prague, vol. 12, pp. 90–100 (2005) ISSN 1210-9479
7. Martin, J.: Principles of Object-Oriented Methods. Prentice-Hall, Englewood Cliffs (1993)
8. Martin, W.J.: The Global Information Society. Aslib Hober, Aldershot (1995)
9. PMBOK 1996. A Guide to the Project Management Body of Knowledge [PDF manual]. Material PMI (1996), http://www.pmi.org
10. Polak, J., Merunka, V., Carda, A.: The art of the system proposal, Grada, Prague (2003) ISBN 80-247-0424-02
11. Managing succesfull projects with PRINCE2 [PDF manual]. London: Stationery office (2002) ISBN 011-330-8914
12. Repa, V.: The analysis and proposal of information systems. Ekopress, Prague (1999) ISBN 80-86119-13-0
13. Repa, V.: Process management and modelling. Grada, a.s. Prague (2006) ISBN 80-247-1281-4
14. Svata, V.: Project management in the circumstances of the ERP systems. In: Oeconomica, Prague (2007) ISBN 978-80-245-11832
15. Talasova, J.: Fuzzy methods of multicriteria evaluation and decision making. In: UP Olomouc, Faculty of natural sciences, Olomouc (2003) ISBN 80-244-0614-4

Chapter 21
The Importance of Virtual Community for Enterprise Functioning – The Case Study of P.P.H.U. Jedroch Partnership Company

Miroslaw Moroz

Abstract. The ease of interaction among the Internet users is a very important feature as far as business is concerned. A group of Internauts can comment a given event, upload their own materials, share their knowledge, make opinions on products. The aim of this article is to present organizational and managing consequences of Internauts' opinions expressed within the framework of virtual communities. Basing on the case study of Jedroch partnership company, a range and character of enterprise reactions upon opinions expressed by current and potential users of company products in the Internet forums was analysed.

21.1 Introduction

The economic implementation of the Internet has changed the way of running business. Introducing the Internet to everyday company's activities is connected with changes in the information, communication, distributive and transaction domain. On-line advertisements, various synchronic and asynchronic forms of communication with the client, entering the new markets thanks to the Internet sale channel are the most common examples of implementing information and telecommunication technologies.

Simultaneously, the Internet is a great tool, not only in hands of entrepreneurs, but also in hand of consumers. Due to the development of the Internet services, Internauts were given tools that allow them to articulate their views upon an unprecedented scale; in a geographic sense as well as in a quantitative sense, at minimal costs. Various kinds of information exchange forums, social media and other forms of expression are taken into account. Internauts, to a large degree, use the Internet for 'horizontal' contacts among themselves. As far as business point of view

Miroslaw Moroz
Department of Economics and Organization of Enterprise, Wroclaw University of Economics
e-mail: miroslaw.moroz@ue.wroc.pl

E. Tkacz and A. Kapczynski (Eds.): Internet – Technical Development and Appl., AISC 64, pp. 185–192.

is concerned, a virtual community is born - a relatively homogenous group, taking interests, needs that may be interested in a given brand products, into account. This group may stand as a point of reference for its members, while doing shopping. As the research shows, expressing one's own opinion has its consequences in purchase decisions of other people [1, 2]. The same way virtual community linked to a given brand or company can influence the internal decisions in enterprise in a considerable way.

The aim of this article is to present organizational and managing consequences, getting to know and recognizing Internauts' opinions as righteous, being expressed within the framework of virtual communities. Basing on the example of P.P.H.U. Jedroch partnership company, a range and character of enterprise reactions upon opinions expressed by current and potential users of company products in the Internet forums was analysed.

21.2 The Role of Virtual Community in Marketing

A modern enterprise communicates with the market through a number of channels. Apart from traditional forms of advertisement and promotion, since the 90' of the 20th century, there have appeared interactive forms connected with the Internet.

The Internet has an influence on the very idea of marketing as philosophy of acting, as well as marketing tools. In the first instance, it all results from the fact that implementing the Internet deepen the marketing orientation of enterprise, and also because ICT assure greater individualization of marketing activity. Intensified product adjustment to clients needs is due to product configuration possibilities through the Internet. An Internet user using on-line order, does not have a standard product, but a specially tailored one to his or her needs. Using this type of activities of web, on a large scale, allows to the so called mass customisation.

As far as the second sphere of marketing tools is concerned, the Internet offers the possibility of informing clients about products or about a given company. The same way it influences the following sale activation instruments, advertising, sale promotion and public relations [3]. One can enumerate many e-marketing instruments on this level; such as: banner, pop-up window, interstitial, toplayer, brandmark. Being placed on a top position in search engines, due to firm's website optimization, or using the paid results in the form of sponsored links and contextual advertising, is the crucial part of e-marketing. Initiating and direct addressing the marketing message towards Internauts is the common denominator for the aforementioned marketing tools. The above presented forms of e-marketing are characterized by decreasing effectiveness (drop of CTR indicator from 0,33 % in 2004 to 0.19 % in 2008 [4]), as a result of Internauts' sense of boredom on the one hand, and some degree of advertisement invasiveness on the other hand.

Recently there have appeared less invasive forms of building enterprise reputation. Opinion exchange among Internauts within virtual communities is in question here. A virtual community can be defined as a group of people sharing common interests, contacting each other electronically [5]. It is the Internet that enabled people to communicate in a way that was not known before. Internauts sharing common interests can exchange opinions in a convenient way as far as place and time are concerned. Contemporary solutions existing in virtual communities give tools of expressing one's own opinion that is convenient in use. In this context one can enumerate [6]: discussion groups (Usenet), mailing lists, discussion forums, social media, blogs, communicators and on-line chats. There exist huge tool diversity that may serve in expressing one's own opinion. As far as forms of expression are concerned, the creator has choice - he or she can insert textual entry, a photo, an audio or video file. Integration of the above mentioned forms is possible.

The change in message sender is the most important change in relation to the earlier presented marketing tools. Virtual community members are themselves authors of expressed opinions and other members of a given community are addressees. Other words - communication upon a product of a given company is between two or more Internauts, beyond enterprise marketing team. Looking from business perspective, a virtual community is perceived as shopping reference group [2].

The phenomenon of recommending companies' products have been known for ages. However, new dimension is being born in the Internet era, due to the number of people and ease of opinion exchange. The number of people is a derivative of the global range of the Internet and the possibility of searching out such a circle that, for instance, exploit a given model of the car, of a given year and equipment. Most probably, it would be hard to find people having such knowledge and experience about the aforementioned product in traditional embedding. It is possible in virtual space and conditioned only by eagerness of sharing his or her knowledge by a given Internet user. As the research shows, such will exists, and the motivators spectrum is wide enough so that, practically, every community member has got his or her own reason for original signalling their presence on the virtual community forum [7]. The ease of expressing one's own opinions results from an intuitive way of using the Internet applications. Additionally, a sense of anonymity has an influence on the subjective ease of expression - the majority of Internauts chooses nick under which is present on various information exchange forums.

Factors creating the meaning of virtual community as a place for making purchase decisions are based on:

- A sense of independence
- Unique characteristics of information generated by Internauts themselves
- Dynamism and interaction

The first basic advantage in making decision is a non-commercial origin and character of the majority of virtual communities. Unlike information prepared by enterprise that has a clear aim in presenting a product in a positive way, opinions expressed by other Internauts are perceived as objective and not infected by a more or less visible advertisement context. Research in the USA shows that Internauts trust and appreciate other Internauts' opinions (even unknown) more than those expressed by experts [1, 8]. The only condition for such a situation is rich opinion base, at least, four opinions [1]. This advantage, however, was noticed by some public relations agencies that used the power and credibility of Internauts' opinions, creating positive opinions from existing clients [9].

The character of information appearing on the Internet forums is the second factor. If a given person seeks practical and experience-based knowledge about functioning of a given product, the only commonly accessible possibility is having other users' opinions. In this sense, virtual communities are characterized by a unique range of information - a precise and very practical one. In majority of cases, these are solid opinions based on users' experience and generated to make other people life easier. As the USA research shows, positive opinions factor exceeds 40% of the total number of opinions about the product and only 27% are negative [10]. It does not exclude, however, being careful in interpretation of entries and rejecting extreme opinions.

The third factor is connected with the fact that content generated by the Internet user grow up dynamically. Contrary to traditional media, an opinion entry posted by a given Internet user, gains new content via comments and discussions. Posting a given piece of information does not end the edition process and is its beginning in fact. It can be viewed in plus as far as making purchase decision is concerned; as the knowledge about a given product constantly grows bigger.

21.3 Organizational and Managing P.P.H.U. Jedroch Reactions upon Opinions Expressed by Internauts

Producing prams is the basic activity of P.P.H.U. Jedroch - partnership company. Because of its essence, such a product undergoes especially strict evaluation of those interested to buy - as it is all connected with safety and comfort of a small child. Material quality, as well as functional technical solutions (the size of the pram, type of tyres, etc.) are important in this business.

P.P.H.U. Jedroch - partnership company has been on the market for 36 years and it has constantly improved the pram production. It currently offers seven types of prams: Betina, Bartatina, Vedi, Sato, Fyn, Pati and Viki [11]. The following products differ from each other as far as design and technical parameters are concerned. Adjustable handle, pumped tyres and the possibility of mounting child safety car chair are the main differences. Additional accessories, such as: an umbrella, a rain cover, a diaper bag and a sleeping-bag are offered along with the product. This company got PN/EN 1888 certificates 'accessories for children, prams, safety requirements and research methods'.

P.P.H.U. Jedroch - partnership company offers prams on the Polish market, as well as it exports its own products to the Czech Republic, Denmark, Norway, Finland, Russia and Germany. The company also began producing under foreign brand in Germany. The number of employed ranges up to 80 people and the income for 2008 circles around 5 million Euro. Therefore P.P.H.U. Jedroch - partnership company meets requirements for medium-size company.

Information generated by Internauts, and referring to enterprise products, play a crucial role in functioning of a given company. The above ascertainment is especially true for the Jedroch company because of specific character of the product. Prams are one of the basic products for a small child and are the subject of profound estimation of current or future parents. Virtual community grouping Internauts interested in subjects connected with children is the proper place for collecting opinions of real advantages and disadvantages of prams. Therefore, the enterprise took the existence of virtual communities into account. There were also specified undertakings due to posted opinions.

As far as the organizational point of view is concerned, Internauts' opinions referring to company products are monitored and analysed on the high level of managing the company. The forum is monitored by the managing director and sale coordinator of the company for the home market. It is the sale coordinator for home market on whom there is competence of monitoring and transferring Internauts' opinions to other enterprise organizational cells (further in the text: about coordinator). Such a division of responsibility is an expression of a great attention, that is paid to opinions placed on the Internet forum. It causes particular perturbations that result from a great number of coordinator duties, on the other hand. Particular cases show that answers to Internauts' doubts, are formulated, on the behalf of the company, with 24 hour delay. This is violation of the so called netiquette - a set of rules of the proper behaviour in virtual space.

The coordinator monitors the forum several times in a month. Frequency of becoming familiar with Internauts' opinions is the function of the workload of remaining, vast range of duties. Due to the lack of time, the coordinator does not take part in discussions run by Internauts. However, when there are reproaches on a charge of the company's products, or there appears a dissatisfied customer, the coordinator tries to explain doubts and indicate the real pram parameters. Frequently, it turns out that the customer's dissatisfaction results from incomplete information in the shop, and only a well-grounded information is enough in such a case. The average time of response also exceeds 24 hours and lasts up to several days.

P.P.H.U. Jedroch - partnership company does not inspire utterances on the forum, as it believes that basing on current experience, articulated opinions are solid and lacking spiteful remarks. The company did not employ, and there no such plans to employ Public Relations agency that is an expert in Word-of-Mouth Marketing. The coordinator, basing on long-term observations, estimates Internauts opinions, expressing their point of view upon Jedroch company prams, in a positive way. Women - young mothers are 95% of discussion forums participants. Women concern about safety and comfort of their own children causes that responses are exceptionally solid as far as advantages and disadvantages of Jedo prams are concerned. There appear practical suggestions for usage various models of prams, especially helpful for people seeking information and aiming to buy the proper pram. Author's own observations only confirm that forum dedicated to prams is exceptionally constructive and lacking spiteful remarks, provocation and eagerness to show off from the side of Internauts.

There was also no observation in Jedroch company revealing competitors as those who, under disguise of Internauts, could run a false or groundless critique.

As far as managing point of view is concerned, the existence of exchange information forum makes the decision making process efficient. Introducing FYN 4 model is an example. It was created on the basis of Internauts 'ifs and buts' as for polyurethane foam tires. Since January 2009 FYN 4 model has been introduced with inflated wheel tires and light aluminium construction. Introducing this model caused a considerable decrease of company's income, as well as it contributed to decrease in employment. Internauts' suggestions are also helpful in other parameters as far as designing prams is concerned. External dimensions of the pram is a very important construction aspect. Internauts' suggestions were also taken into consideration in this case; so that the pram could fit into the smallest elevator.

Basing on Internauts' utterances, the www site of the company was modified. Changes were connected with the way of searching several models, as well as an introduction of the possibility of the direct comparison of all the pram models as for important parameters in the eyes of the potential user - weight, dimensions, equipment, etc.

P.P.H.U. Jedroch - partnership company gains three types of benefits as a result of functioning of virtual communities interested in prams:

- financial
- marketing
- developmental

The economic benefits are based on opinions of previous Jedroch company pram users. Complimentary opinions made many future mothers buy prams produced by this company. The income increased 20% in 2008 in comparison to the previous year. The analysis in Jedroch company shows that positive Internauts' opinions contributed to the half of the aforementioned sale increase. This fact proves the meaning of virtual community for functioning of the enterprise. The company 'did not spend a penny' for such a form of promotion. However, one has to highlight that a similar effect can be obtained only by those enterprises which products are solidly done and meet real customers' needs.

Marketing benefits are connected with promoting Jedo brand among existing and potential product users. The awareness of the brand grows in the target group and this target group - potential mothers get familiar with the company products without a sense of invasiveness of the commercial advertisement sort, as well as it has the possibility of discussing practical aspects of pram usage of different companies. It is conducive to making a more objective purchase decisions.

Developmental benefits are connected with Internauts' suggestions that are taken into account while designing prams. Internauts, as everyday company product users are the source of many interesting and legitimate ideas improving the product. The earlier mentioned examples as for the changes in tyres of physical dimensions show the potential of the virtual community also in the area of development and research.

The above types of benefits are mutually connected and decide about synergy effects resulting from the existence of virtual community. It is worth highlighting that, in the coordinator's opinion, there are no disadvantages of information exchange forums accompanying advantages. There weren't any disadvantages, at least, there were not noticed ones in Jedroch company.

21.4 Conclusions

Virtual communities are an indication of using unprecedented possibilities of entering various relations and interactions through the Internet. Coming into being such types of communities has, generally, a non-commercial origin and 'the rank-and-file' character. It does not cross out, however business dimension of the virtual community. Because of its characteristics, opinions formed by community members can help or hamper enterprise activity. Virtual communities play an important role in enterprise marketing from this perspective, however they are beyond control of marketing department.

The role of virtual community as the shopping reference group will grow together with entering new phases of the so called 'Y' generation, development of intuitive applications for self-articulating one's own opinions and growing popularity of various kinds of social media. The analysed example of the Jedroch company proves that virtual community connected with the company product is an essential component that has to be taken into account in decision processes. Basing on, valuable for customer, product, the Jedroch company is satisfied with positive relations of previous users, that it, in turn, is transformed into economic, marketing and developmental benefits. Specific character of products of the analysed company causes that the target group is constructed in 95% of mother or future mothers; that are concerned about safety and comfort of their children. Therefore, the exchange information forum has such a constructive and serious dimension. Not every branch will be in a similar situation. It does not mean that Jedo prams are the subject of critique, on the other hand. As it is stated by representatives of the company, women' estimations are exceptionally accurate and solid.

Analysing the situation of interactions among Jedroch company and existing virtual community, one may point at two activities (measures) that would enlarge synergy effect of mutual interaction. Firstly, the coordinator's work overload is more than vivid. This function imposes responsibility for area of communication and analysis of virtual community. It results in delays in responses and/or low frequency of monitoring of forums. It seems to be accurate to lessen his or her remaining burden. Secondly, it would be worth implementing elements that help creating community (forum or chat) on the company's www site. The aim would be to enrich the message of the company's www site with the customers' opinions.

References

1. eMarketer: Online Reviews Sway Shoppers,
   http://www.emarketer.com/Article.aspx?id=1006404
2. Pentina, I., Prybutok, V.R., Zhang, X.: The role of virtual communities as shopping reference groups. Journal of Electronic Commerce Research 9(2), 116 (2008)
3. Wielki, J.: Elektroniczny marketing poprzez Internet, p. 110. PWN, Warszawa-Wroclaw (2000)
4. eMarketer: The Latest Ad Click Count,
   http://www.emarketer.com/Article.aspx?id=1006969
5. Hye-Shin, K., Byoungho, J.: Exploratory study of virtual communities of apparel retailers. Journal of Fashion Marketing and Management 10(1), 41 (2006)
6. Mazurek, G.: blogi i wirtualne spolecznosci - wykorzystanie w marketingu, Wolters Kluwer Business, Krakow, p. 104 (2008)
7. Wang, Y., Fesenmaier, D.R.: Towards understanding members' general participation in and active contribution to an online travel community. Tourism management 25, 717 (2004)
8. eMarketer: The Growing Influence of Online Social Shoppers,
   http://www.emarketer.com/Article.aspx?id=1006146&src=article1_newsltr
9. Miaczynski, P., Grynkiewicz, T.: Nie myslcie, ze to jakas sciema, Gazeta Wyborcza, p. 9 (03.03.2009)
10. eMarketer: Who's Afraid of the Big Bad Bloggers?,
    http://www.emarketer.com/Article.aspx?R=1007054
11. Jedo Partnership Company, http://www.jedo.pl/2009/index.html

Part III
Information Security in Distributed Computer Systems

Chapter 22
Quantum Cryptography: Opportunities and Challenges

Andrzej Grzywak and George Pilch-Kowalczyk

Abstract. Quantum information processing is a new and dynamic research field at the crossroads of quantum physics and computer science. Once a quantum computer becomes reality, it will possess massive parallel processing capabilities. Although this goal is still quite distant, certain limited applications have been developed. One of them is quantum cryptography, exploiting Heisenberg's uncertainty principle to allow two remote parties to exchange a cryptographic key without a possibility for a third party to intercept the key during its exchange between the sender and the recipient. Current popular exchange of keys using public key cryptography suffers from two major flaws. First, it is vulnerable to technological progress. The development of the first quantum computer will consequently make the exchange of a key with public key algorithms insecure. The second flaw is the fact that public key cryptography is vulnerable to progress in mathematics. These threats simply mean that public key cryptography cannot guarantee future-proof key distribution. Quantum cryptography solves the key distribution problem by allowing the exchange of a cryptographic key between two remote parties with absolute security, guaranteed by the laws of physics. Mechanics of this exchange has been described in the paper. The quantum cryptography system is very promising and advancements are being made to improve upon the technology, most notably a wireless implementation, but it is still susceptible to hacker attacks and has transmission distance and encryption rate limitations. This paper will discuss the flaws of quantum cryptographic systems along with the plans for enhancing current quantum cryptographic systems.

22.1 Introduction

Information security is one of essential issues in contemporary computer systems. This paper deals with secure transmission of information based on principles of

Andrzej Grzywak · George Pilch-Kowalczyk
Academy of Business, Dabrowa Gornicza, Poland
e-mail: agrzywak@wsb.edu.pl, school@ninta.com

E. Tkacz and A. Kapczynski (Eds.): Internet – Technical Development and Appl., AISC 64, pp. 195–215.
springerlink.com

encryption. Encryption of information long precedes computer age - vivid example could be Caesar's cipher used some two thousand years ago. Progress in information security is shown on Fig. 22.1; though for last 40 years it is related to computer systems. Modern networks generally rely on one of two basic cryptographic techniques to ensure the confidentiality and integrity of traffic carried across the network: symmetric (secret) key and asymmetric (public) key. Indeed today's best systems generally employ both, using public key systems for authentication and establishment of secret "session" keys, and then protecting all or part of a traffic flow with these session keys.

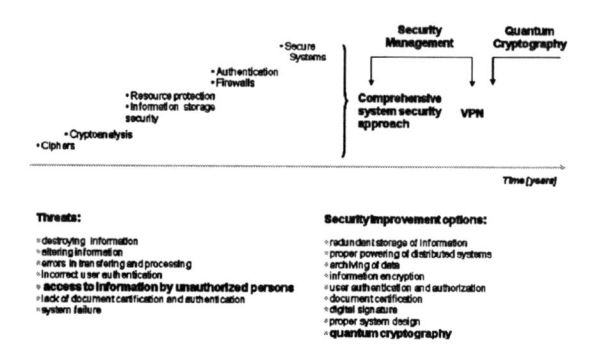

Fig. 22.1 Progress in information security solutions

Early ciphers used in computer systems were proposed by IBM, and approved by NBS (National Bureau of Standards) as Data Encryption Standard (DES). DES has been using symmetric keys for both encryption and decryption. This concept is depicted on Fig. 22.2, where information is sent as cipher text using unsecured channel, and a secure channel is used to send the key. In recent years, the cipher has been superseded by the Advanced Encryption Standard (AES). Secure transfer of encryption keys remains one of essential problems in modern cryptography. In pursuit of more adequate method, RSA algorithm for public-key cryptography was developed in late 70-ties. RSA involves a public key and a private key. For public key cryptographic systems, it is no longer necessary to exchange keys over a secure channel. Instead, both sides create their own individual encryption/decryption key pairs. Then they both keep their decryption keys secret from everyone, and "publish" or

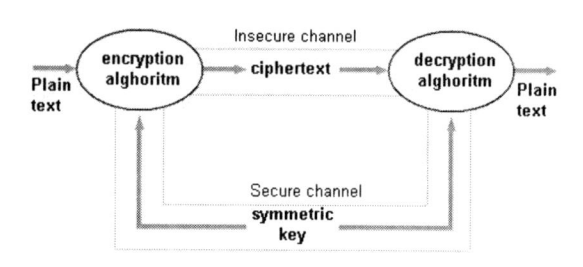

Fig. 22.2 DES information channel

publicly broadcast their encryption keys. The public key can be known to everyone and is used for encrypting messages. Messages encrypted with the public key can only be decrypted using the private key. In practical implementations, asymmetrical algorithms are used not so much for encryption, because of their slowness, but rather for distribution of session keys for symmetrical cryptosystems, as it has been implemented in widely used SSL encryption.

The security of RSA is actually based on the factorization of large integers. A key should therefore be large enough that a brute force attack would take too long to execute. Outline of RSA algorithm is as follows:

- Pick two large prime numbers p and q and calculate the product $N = pq$, $f = (p - 1)(q - 1)$
- Choose a number c that is co-prime with f
- Find a number d to satisfy $cd = 1 \ mod \ f$, using a method such as Euclid's algorithm
- Using your plaintext, a, the cipher text is encoded as $b = a^c \ mod \ N$
- To retrieve the plaintext, $a = b^d \ mod \ N$
- The numbers N and c are made public, so anyone can encrypt information, but only someone with d can retrieve the plaintext

Breaking RSA encryption means given N compute p and q, taking exponential length of time along with increase of prime numbers used in encryption.

One weakness of public key cryptography comes from a flaw, that it is vulnerable to technological progress. Although initially RSA encryption utilized much shorter keys, as of 2002 a key length of 1024 bits was generally considered the minimum necessary for the RSA encryption algorithm due to technological progress in computer efficiency. Conversely, commonly adopted key length for symmetrical encryption such as AES is 128 bit (equivalent to 3072-bit RSA keys) with possibility of increase to 256 bit [1].

Another weakness lies in vulnerability to progress in mathematics. The ominous example is development of Shor's quantum factoring algorithm in 1994, a polynomial algorithm allowing fast factorization of integers with a quantum computer.

Shor's algorithm consists of two parts [2]: A reduction, which can be done on a classical computer, of the factoring problem to the problem of order-finding. A quantum algorithm to solve the order-finding problem.

On a quantum computer, to factor an integer N, Shor's algorithm takes polynomial time in ln N, specifically $O((\ln N)^3)$. This is exponentially faster than the best-known classical factoring algorithm, the general number field sieve, which works in sub-exponential time - about.

Although it was designed for future quantum computers, it demonstrates that once quantum computers became reality, classical cryptography, as we know it, no longer would be secure.

Quantum computers will perform computations at the atomic scale, exploiting rules of quantum mechanics to process information in ways that are impossible on a standard computer. The basic principle behind quantum computation is that quantum properties such as superposition, entanglement, uncertainty, and interference can be used to represent data and perform operations on these data. They would solve certain specific problems, such as factoring integers, dramatically faster compared to silicon computers.

The fundamental feature of a quantum computer is that it uses qubits instead of bits, but unlike classical bits, qubits can exist simultaneously as 0 and 1, with the probability for each state given by a numerical coefficient. A quantum computer promises to be immensely powerful because it can be in multiple states at once-a phenomenon called superposition–and because it can act on all its possible states simultaneously.

22.2 Quantum Key Distribution

Quantum cryptography - more precisely Quantum Key Distribution (QKD) - is already making inroads towards practical use outside of laboratories. Use of quantum computers on one hand creates threat to security of classical cryptography, however on the other hand quantum computing opens a road to even more secure distribution of secret keys.

Quantum information is stored as the state of atomic or sub-atomic particles. A qubit is an elementary unit of quantum information.

Some of the physical realizations of a qubit can be:

- **An electron.** The information is encoded as the spin of the electron.
- **A photon.** The information is encoded as the photon polarization.
- **Quantum dots.** Small devices that contain a tiny droplet of free electrons and many others.

The Stern-Gerlach experiment often used to illustrate basic principles of quantum mechanics was performed in 1922 by Otto Stern and Walther Gerlach on the deflection of particles. They demonstrated that electrons are spin 1/2 particles. These have only two possible spin angular momentum values, called spin-up and spin-down (Fig. 22.3). Spin of electrons can be considered as an equivalent to binary system. In the experiment spin-up could mean logical 0, and spin-down could mean logical 1. A significant step in utilization of quantum mechanics laws has been use of light in quantum cryptography. In telecommunication networks, light is routinely used to exchange information in a form of light pulses, typically containing millions of particles of light, called photons. In quantum cryptography it is reduced to a single photon. Polarization of light is the direction of oscillation of the electromagnetic field associated with its wave. Fig. 22.4 demonstrates a principle of horizontal and vertical polarization of light with the use of crystals. We can therefore define vertically polarized light (Fig. 22.5), as well as a vertically polarized photon (Fig. 22.6). Similarly light can be polarized horizontally. Light can be diagonally polarized at 45° (Fig. 22.7). If the rotating electric field vector inscribes a circle, we call this polarization the right- or left-circular. The quantum cryptographic protocols will use some encoding scheme which associates the bits 0 and 1 with distinct

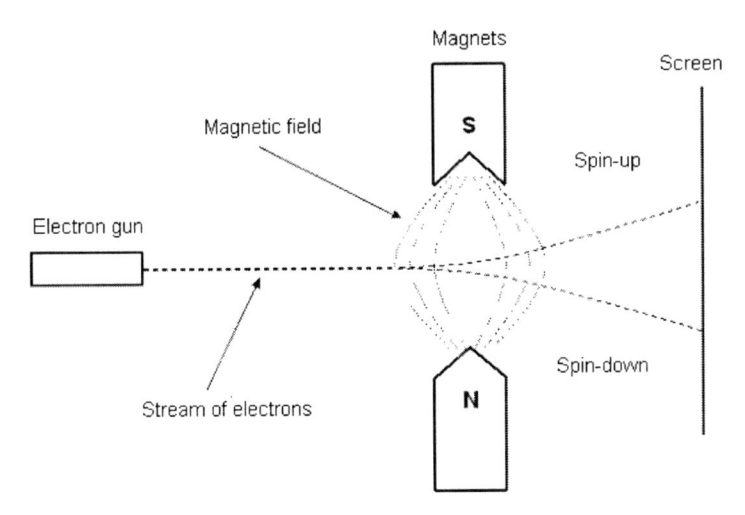

Fig. 22.3 Stern-Gerlach experiment

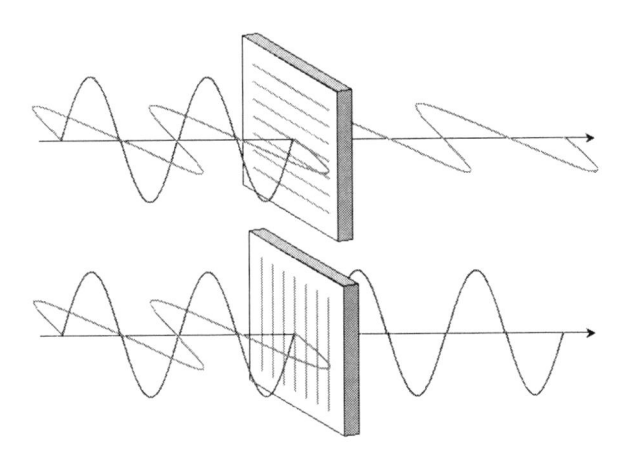

Fig. 22.4 Polarization of light

quantum states. Such an association is called a quantum alphabet (Fig. 22.8). Quantum cryptography utilizes the Heisenberg uncertainty principle of quantum mechanics, leading to the observation that one cannot take a measurement without perturbing the system. Communicating parties can check whether someone "was listening": they simply compare a randomly chosen subset of their data using a public channel, however they could discover any eavesdropper only after they have exchanged their message. It would of course be much better to ensure their privacy in advance and not afterwards. It however does not matter if we use the quantum channel only to transmit a random sequence of bits that is a key. Now, if the key is unperturbed, then quantum physics guarantees that no one has gotten any information about this key by measuring (eavesdropping) the quantum communication

Fig. 22.5 Light polarized
vertically

Fig. 22.6 Vertically polarized
photon

Fig. 22.7 Photon diagonally
polarized at 45°

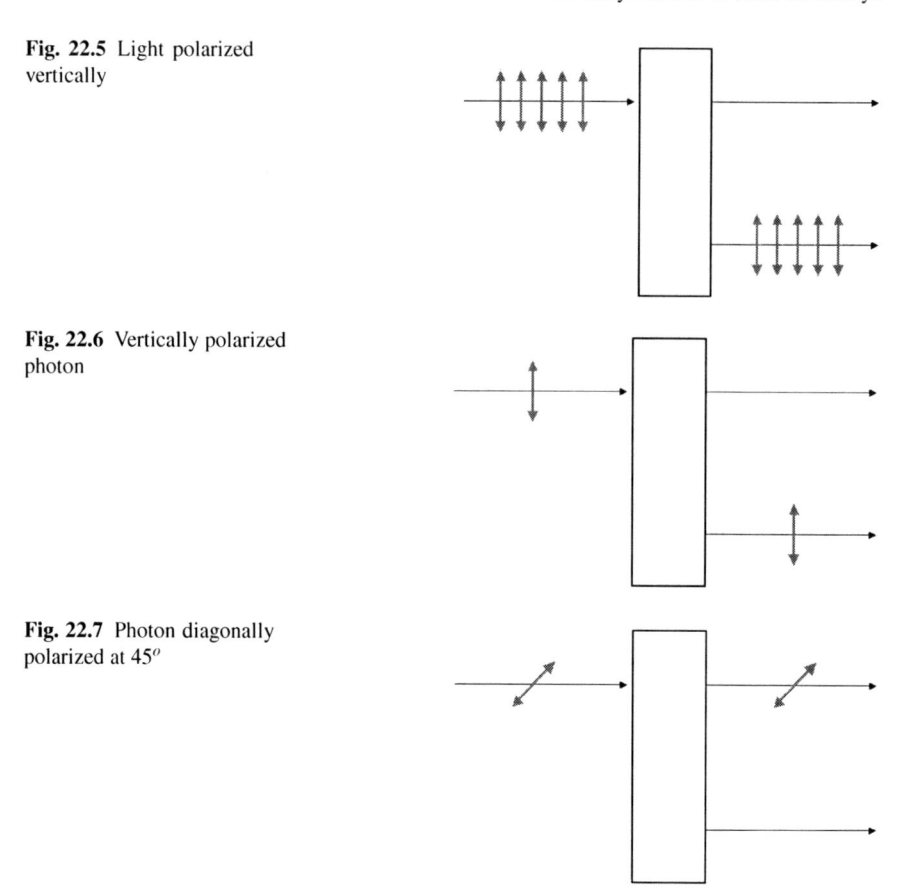

Basis	Representation	Random Bit 0	Random Bit 1
Rectilinear	+	↑	→
Diagonal	X	↗	↘
Circular	O	↺	↻

Fig. 22.8 Quantum alphabet

channel. In this case, the key can be safely used to encode messages. If, on the other
hand, the key turns out to be perturbed, then it can be simply disregarded. Since the
key does not contain any information, there is no loss of any kind.

Quantum key distribution therefore employs two separate channels (see Fig. 22.9). Quantum channel is used for transmission of quantum key material by means of photons. The other, public (classical) channel carries all message traffic, including the cryptographic protocols, encrypted user traffic, etc. [3] Quantum information has special properties: the state of a quantum system cannot be measured or copied without disturbing it quantum state can be entangled, two systems have a definite state though neither has a state of its own superposition - we cannot reliably distinguish non-orthogonal states of a quantum system.

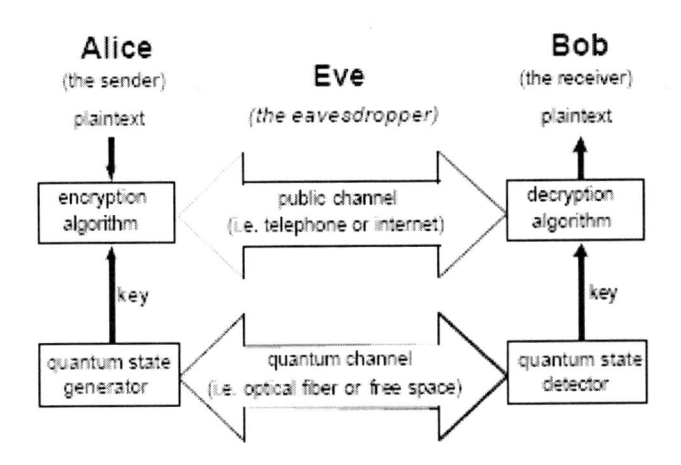

Fig. 22.9 Quantum Key Distribution

These properties create major difference between quantum cryptography technology and traditional cryptographic technology. Quantum cryptography relies on the laws of quantum mechanics to provide a secure system, while traditional systems rely on the computational difficulty of the encryption methods used to provide a secure system.

22.3 Quantum Cryptography Protocols

22.3.1 The BB84 Quantum Cryptographic Protocol

Creation of a quantum alphabet was a basis for development of quantum cryptographic communication protocols. The first quantum cryptographic communication protocol, called BB84, was invented in 1984 by Bennett and Brassard [16]. This protocol has been experimentally demonstrated to work for a transmission over fiber optic cable [22], and also over free space [23]. Although new emerging protocols, apparently more efficient, are gaining popularity nowadays, BB84 still remains in use.

The BB84 protocol utilizes any two incompatible orthogonal quantum alphabets, such as depicted on Fig. 8. Bennett and Brassard note that, if Alice were to use only one specific orthogonal quantum alphabet for her communication to Bob, then Eve's eavesdropping could go undetected. To assure the detection of Eve's eavesdropping, Bennett and Brassard require Alice and Bob to communicate in two stages [9] , the first stage over a one-way quantum communication channel from Alice to Bob, the second stage over a two-way public communication channel(Fig. 22.10).

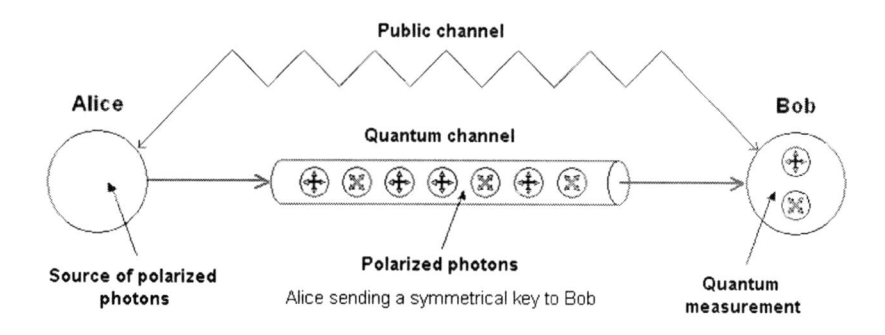

Fig. 22.10 Quantum key distribution employing BB84 protocol

In the first stage, Alice is required, each time she transmits a single bit, to use randomly with equal probability one of the two orthogonal alphabets A_+ or A_x. Since no measurement operator of A_+ is compatible with any measurement operator of A_x, it follows from the Heisenberg uncertainty principle that no one, not even Bob or Eve, can receive Alice's transmission with accuracy greater than 75%. Because Bob's and Eve's choice of measurement operators are stochastically independent of each other and of Alice's choice of alphabet, Eve's eavesdropping has an immediate and detectable impact on Bob's received bits. In second stage, Alice and Bob communicate over a public channel to check for Eve's presence by analyzing Bob's error rate. First phase of this stage is aimed to eliminating the bit locations at which error could have occurred without Eves eavesdropping. Bob begins by publicly communicating to Alice which measurement operators he used for each of the received bits. Alice then in turn publicly communicates to Bob which of his measurement operator choices were correct. After this two way communication, Alice and Bob delete the bits corresponding to the incompatible measurement choices to produce shorter sequences of bits which we call respectively Alice's raw key and Bob's raw key. If there is no intrusion, then Alice's and Bob's raw keys will be in total agreement.

Second phase deals with Eve's presence. For each bit transmitted by Alice, we assume that Eve performs one of two actions, opaque eavesdropping with probability λ, $0 \leq \lambda \leq 1$, or no eavesdropping with probability 1 - λ. Thus, if lambda = 1,

Eve is eavesdropping on each transmitted bit; and if $\lambda = 0$, Eve is not eavesdropping at all.

If Eve has been at work, then corresponding bits of Alice's and Bob's raw keys will not agree with probability $0 * (1 - \lambda) + V * \lambda = V * \lambda$.

Ideally, in the absence of noise, any discrepancy between Alice's and Bob's raw keys is proof of Eve's intrusion. So to detect Eve, Alice and Bob select a publicly agreed upon random subset of **m** bit locations in the raw key, and publicly compare corresponding bits, making sure to discard from raw key each bit as it is revealed.

Should at least one comparison reveal an inconsistency, then Eve's eavesdropping has been detected, in which case Alice and Bob return to stage 1 and start over. On the other hand, if no inconsistencies are uncovered, then the probability that Eve escapes detection is [9]:

$$P_{false} = (1 - V\lambda)^m$$

For example, if Eve is eavesdropping on each transmitted bit ($\lambda = 1$) and length of key subset is just 200, then

$$P_{false} = (3/4)^{200} = 10^{-25}$$

Thus, as P_{false} is sufficiently small (usually no more than 10^{-9}), Alice and Bob agree that Eve has not eavesdropped, and accordingly adopt the remnant raw key as their final secret key.

In reality, noise unavoidable due to technical, material reasons, introduces errors still with the assumption that all errors in raw key are caused by Eve. Comparison in the subset **m** will then reveal errors at an estimate **R** of the error-rate. If **R** exceeds a certain threshold R_{max}, then Alice and Bob return to stage **1** to start over, otherwise revealed error bits will be removed from the key. Then a process of reconciliation called also a key distillation will follow.

First step in this process employs a classical error correction protocol, to get a shorter key without errors, reducing error rate from few percent to usual 10^{-9}. After error correction, Alice and Bob have identical copies of a reconciled key, but Eve may still have some information about it, therefore it is only partially secret from Eve. The next step is the **privacy amplification**, being the process whereby Alice and Bob reduce Eve's knowledge of their shared bits to an acceptable level. This technique is also sometimes called advantage distillation. Privacy amplification is used to convert the realized reconciled key into a smaller length key through some hashing function chosen at random from a known set of hashing functions.

Finally, in order to prevent certain form of "Man-in-the-middle" attack (Fig. 22.11), there is a need of initial authentication before any exchange of a secret key over a secure communication channel could take place. This can be done in several ways such as for example the way GSM is using; however it may carry some implications on the overall security of the system.

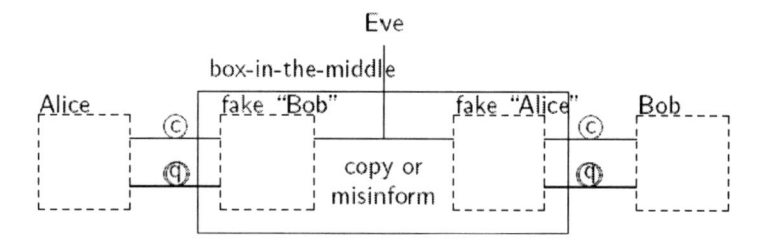

Fig. 22.11 Man-in-the-middle attack

22.3.2 The B92 Quantum Cryptographic Protocol

The B92 protocol is an extension of BB84 which shows how photons with non-orthogonal states can be used to distribute a secret key [24]. As in BB84, Alice and Bob communicate in two stages, the first over a one-way quantum channel, and the second over a two-way public channel. Unlike BB84 which requires two incompatible orthogonal quantum alphabets, B92 requires only a single nonorthogonal quantum alphabet. In the B92 coding scheme, the bit b=0 is encoded by a photon with horizontal polarization and the bit b=1 is encoded by a photon with diagonal polarization at 45°.

In the first stage, Alice is required, each time she transmits a single bit, to use randomly with equal probability either of two nonorthogonal pure states from the alphabet A_θ. Since no measurement can distinguish two non-orthogonal quantum states, it is impossible to identify the bit with certainty. Moreover, any attempt to learn the bit will modify the state in a noticeable way. Bob performs a test which provides him with a conclusive or inconclusive result, using one of many possible measurement strategies, such as suggesting that the measurements will be based on the two incompatible experiments.

Stage 2 for the B92 protocol is similar to that for the BB84 protocol. Alice and Bob use a public channel to inform which bits were identified conclusively, and to compare some of the common bits in order to estimate the error rate. They must accept some small error rate due to imperfections in handling the quantum states. If the estimated error rate exceeds the allowed error rate they return to stage 1 and start over.

22.3.3 EPR Quantum Cryptographic Protocols

Another encoding scheme gaining popularity, also called the Ekert encoding scheme (E91), is similar to BB84, but is based on two photons, called entangled photons[25]. These photon pairs can be created by either Alice, Bob or a third party by splitting a single photon into two, using a laser. After the split, one of the photons is sent by the sender or on behalf of the sender to the receiver while the other photon is kept. The entangled photons follow a principle similar the Heisenberg's Uncertainty Principle where disturbing, monitoring or measuring the state of one entangled photon will disturb the other entangled photon no matter how far apart the entangled paired photons are separated (Fig. 22.12). This property was described as the EPR Paradox (Einstein, Podolsky, Rosen) questioning completeness of the quantum mechanics theory. The EPR quantum protocol is a 3-state protocol that uses Bell's inequality. In the first stage occurring over the quantum channel, for each time slot, a state is randomly selected with equal probability from the set of states. Than an EPR pair is created in the selected state. One photon of the constructed EPR pair is sent to Alice, the other to Bob. Alice and Bob at random with equal probability separately and independently select one of the three measurement operators, and accordingly measure their respective photons. Alice records her measured bit. On the other hand, Bob records the complement of his measured bit to detect the presence or absence of Eve as a hidden variable.

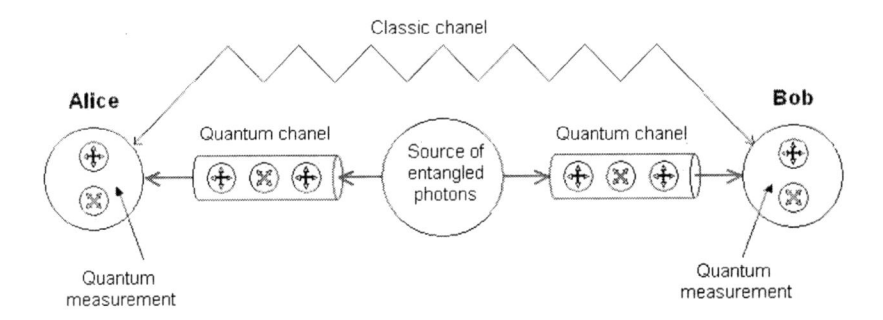

Fig. 22.12 Quantum key distribution employing Ekert protocol

In stage 2 Alice and Bob discuss over a public channel which measurement basis they used for each photon. The two parties then separate the bits of the transmission into two groups called raw key and rejected key. The raw key group contains the bits where Alice and Bob used the same basis for measurement. The rejected key group contains all the other bits. Now, Alice and Bob compare over a public channel their respective rejected key. If their comparison satisfies Bell's inequality then a third party has been detected, then the entire process is repeated. Otherwise the raw key is retained. Unlike the BB84 and B92 protocols, the EPR protocol, instead of discarding rejected key, actually uses it to detect Eve's presence. Alice and Bob now carry on a discussion over a public channel comparing their respective rejected keys to determine whether or not Bell's inequality is satisfied. If it is, Eve's presence is detected. If not, then Eve is absent, and the remainder of the protocol is similar to that of BB84.[9]

22.3.4 Quantum Teleportation

Quantum teleportation in quantum cryptography can be seen as the fully quantum version of the one-time pad [26]. Shannon's work on information theory showed that to achieve perfect secrecy, it is necessary for the key length to be at least as large as the message to be transmitted and only used once (this algorithm is called the One-time pad). In that context, quantum teleportation involves the transfer of an unknown quantum state over an arbitrary spatial distance by exploiting the prearranged entanglement of quantum systems in conjunction with the transmission of a minimal amount of classical information (Fig. 22.13). First, a source of entangled (EPR) particles is prepared. Sender and receiver share each particle from a pair emitted by that source. Second, a Bell-operator measurement is performed at the sender on his EPR particle and the teleportation-target particle, whose quantum state is unknown. Third, the outcome of the Bell measurement is transmitted to the receiver via a classical channel. This is followed by an appropriate unitary operation on the receiver's EPR particle.

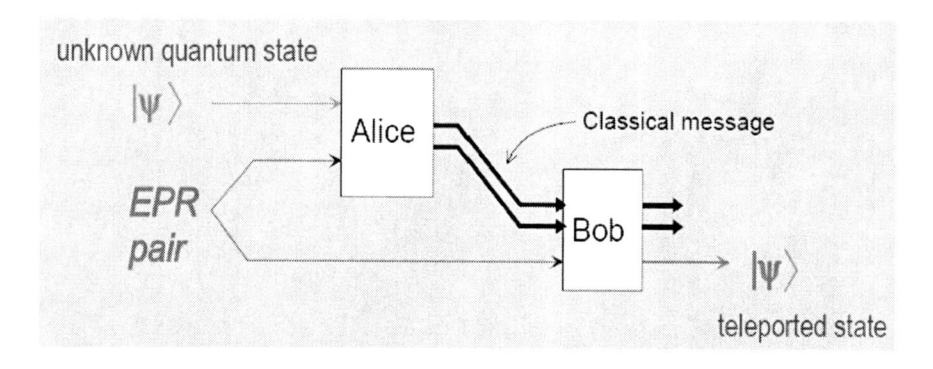

Fig. 22.13 Quantum Teleportation

The name "*teleportation*" is justified by the fact that the unknown state of the transfer-target particle is destroyed at the sender site and instantaneously appears at the receiver site. Actually, the state of the EPR particle at the receiver site becomes its exact replica, but the teleported state is transported between the two sites without transferring the media containing information, therefore it should be immune from eavesdropping.

22.4 Quantum Cryptography in Practice

Today, QKD is no longer confined to laboratories; commercial systems are available, capable of automated continuous operation using available standard telecom fibers. In the SECOQC project of the 6th Framework Programme of the European Community six technologically different systems were operated under realistic assumptions in a QKD network in Vienna in October 2008, providing user level applications with cryptographic keys [27]. Commercial products for point-to-point QKD are today available from at least three small companies (id Quantique, Geneva, SmartQuantum Group, MagiQ Technologies Inc.), supplying products mainly for experimental evaluation. Some major companies are working currently on QKD products as well. Few larger scale experimental projects were designed, such as The DARPA Quantum Network, Swiss voting ballots, The Vienna network, and far advanced projects in Spain and South Africa.

22.4.1 The DARPA Quantum Network

BBN Technologies (Cambridge, MA) operates the world's first quantum cryptographic network, which links several different kinds of QKD systems (Fig. 22.14). Some use off-the-shelf optical lasers and detectors to emit and detect single photons; others use entangled pairs of photons. This DARPA-funded network runs between BBN, Harvard, and Boston University, a city sized schematic designed to test the robustness of such systems in real-world applications [3]. BBN security model is the cryptographic Virtual Private Network (VPN), where existing VPN key agreement primitives are augmented or completely replaced by keys provided by quantum cryptography. The remainder of the VPN construct is left unchanged; see Fig. 22.14. Thus such QKD-secured network is fully compatible with conventional Internet hosts, routers, firewalls, and so on.

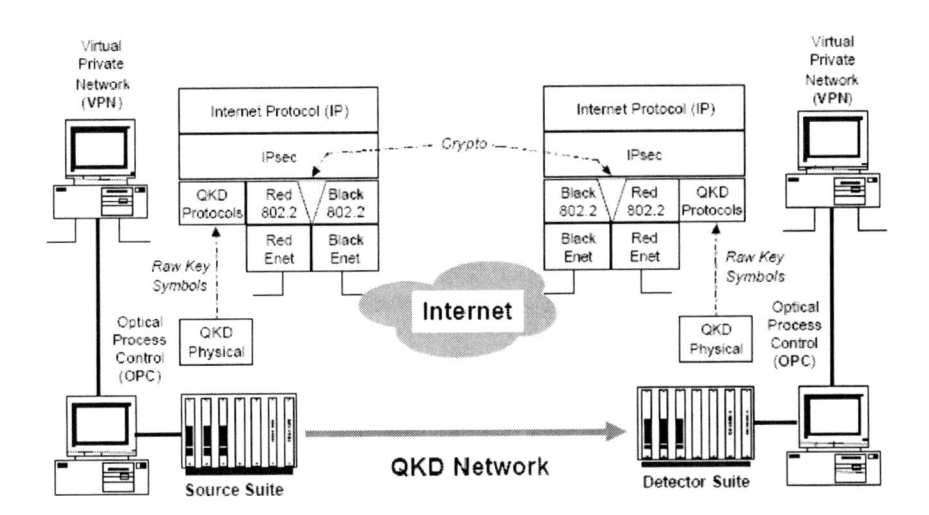

Fig. 22.14 Virtual Private Network (VPN) based on Quantum Key Distribution

22.4.2 The Vienna Network

The network, which is based in Vienna, Austria, was developed under the integrated EU project "Development of a Global Network for Secure Communication Based on Quantum Cryptography" (SECOQC). The Vienna network consists of six nodes and eight intermediary links with distances between 6 and 82 km. There are seven links utilizing commercial standard telecommunication optical fibres and one free-space link. Hardware to the Vienna network was supplied by Toshiba, UK and sites connected in the network were Siemens sites.

22.4.3 Commercial Quantum Products

Swiss company id Quantique is offering a variety of quantum technology products, such as single photon detectors, random number generators as well as quantum key distribution equipment and even complete encryption systems using QKD technology.

An example of Quantum Key Distribution System is Clavis2 that uses a proprietary auto-compensating optical platform, featuring good stability and interference contrast, guaranteeing low quantum bit error rate. Secure key exchange becomes possible up to 100 km. It consists of two stations controlled by one or two external computers. A software suite implements automated hardware operation and complete key distillation. Two quantum cryptography protocols: BB84 and SARG are implemented.

Complete encryption system with QKD is represented by Cerberis (Fig. 22.15), combining QKD solution with few encryption devices implementing AES protocol. The Cerberis solution integrates into existing fiber-optic network infrastructures. [6] Equipment from id Quantique has been successfully used in Geneva electronic voting.

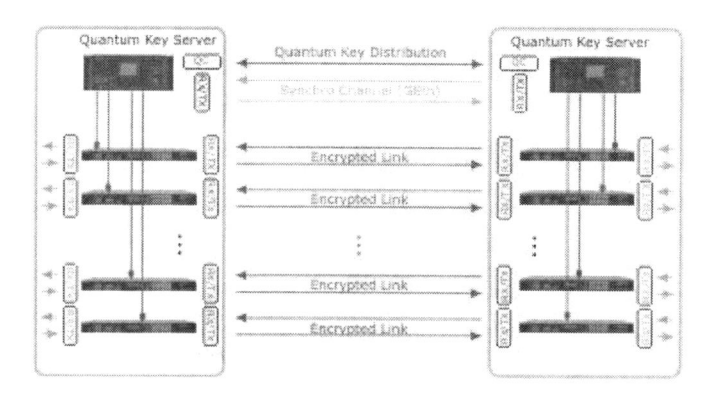

Fig. 22.15 Encryption solution with Cerberis

US based MagicQ, Inc. is another quantum technology vendor. MAGIQ QPN 8505 Security Gateway is a highly-compatible, hardware-based, VPN security solution built on quantum cryptography.[7]. MagiQ QPN solves the problem of refreshing encryption keys regularly as often as 100 times per second by incorporating real-time, continuous, symmetrical quantum key generation based on truly random numbers (Fig. 22.16). MagiQ QPN 8505 comprises of a set of industry standard protocols including BB84, IPSEC based VPN and AES. MagiQ QPNs were implemented in the DARPA network in Boston, MA.

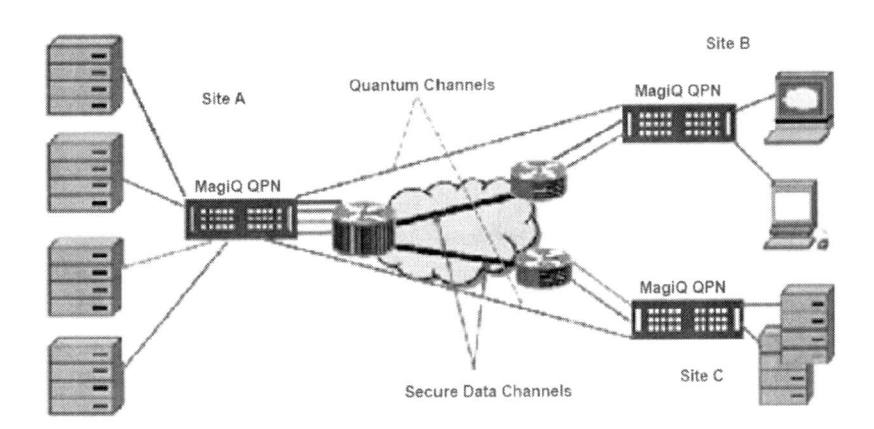

Fig. 22.16 Multi-Site Network Security

22.5 Security of Quantum Cryptography

Quantum cryptography, as introduced, was believed to be a perfectly secure way of communication based on the laws of physics. However, along with theoretical and practical implementation progress many researchers undertook a scrutiny on security of QKD based communication systems, realizing that there are many fine points in the protocol and hardware.

22.5.1 Overall Security of a Communication System

Currently, quantum technology provides a solution to one only component of the secure communication system, namely quantum key distribution (QKD). Such a system can generally be as secure as its weakest component.[5]. From our earlier characterization we know that there two other components required. First, in order to avoid a man-in-the-middle attaca, communicating parties must use authenticated channel; secondly there is a classic channel such as AES to carry encrypted data. There is only one unconditionally secure system, namely so called one-time-pad that requires both parties to use the same random key of length equal to the length of the message and used once only. This requirement becomes impractical for majority of applications, however if the one-time pad is used as the encryption algorithm, then the overall communications system can also be made unconditionally secure.

Another weak component remains anyways i.e. requirement of having an authentic channel. All currently existing authentication schemes which offer unconditional security depend on a pre-established symmetric key. A good example is GSM system, where a Subscriber Identity Module (SIM) contains a 128 bit symmetric key which is shared with the subscriber's network service provider. This key is used in an authentication protocol, one product of which is a new symmetric data encryption key, similar way as QKD systems do.

22.5.2 Attacks on Quantum Cryptosystems

Quantum cryptography systems are vulnerable to a variety of hacker attacks. Three types of attacks relevant to ideal systems could be mentioned: man-in-the-middle (MITM), denial of service (DoS), and large pulse attack.[8].

Man-in-the-middle (MITM) attacks can be performed in two different ways. The first, involves Eve pretending to be "Alice" to Bob and "Bob" to Alice (Fig. 11). Eve would then perform QC with both Alice and Bob at the same time, obtaining two keys, one for Alice and one for Bob. Alice's key would be used to decrypt a message from Alice then reencrypted by Bob's key and vice versa. This type of attack is possible, but preventable by performing some type of identity authentication. The second type of MITM attack comes from the method photons are transmitted. Most of currently implemented systems do not use single photon sources, but rather very weak laser pulses that are small bursts of coherent light. In theory, Eve may be able to split a single proton from the burst without being detected. Eve could then observe the retrieved photons until the basis used to create then is announced.

Denial of service (DoS) attack can be performed in two ways: by compromising the quantum cryptographic hardware or by introducing extra noise into the QC system. Eve the hacker could tamper with the fiber-optic lines, or compromise QC equipment to generate photons at random that are not secure. Excessive noise could cause Alice and Bob to discard a higher number of photons.

Large pulse attack: Eve, the eavesdropper, sends a large pulse of light back to Alice in between Alice's photon transmissions. Regardless of how black Alice's transmitting equipment may be, some light will be reflected back to Eve where she can discover the polarization state of Alice's polarizer because the reflected light will be polarized in the same manner. We need to consider a theoretical/experimental countermeasure against this attack.

22.5.3 Exploiting Physical Imperfections

Let us now consider the security of the non-ideal protocol, taking into account un-avoidable technical imperfections.

A major problem in the implementation of BB84 is generation of a single-photon state. In most experiments, an attenuated coherent laser source is used instead of a perfect single-photon source. All photon sources so far have some probability of multi-photon emission, from which Eve can obtain information by exploiting the so-called photon number splitting (PNS) - see Fig. 22.17. Eve may suppress single-photon signals, and allow passing only those signals that she can split. Since this attack is one of the greatest threats to BB84, protocols with PNS tolerance have been considered. The differential phase shift (DPS)-QKD [17], SARG04 protocol [18], and decoy state method [19], [20], [28] are examples of such protocols or PNS-attack-resistant methods.

Fig. 22.17 PNS attack

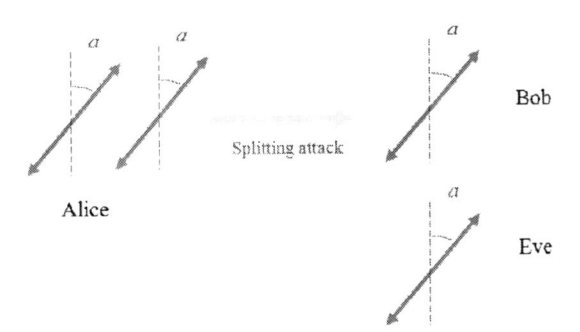

Increased interest in researching quantum attacks gradually evolved to include such imperfections of physical apparatus as faint pulse sources (as opposed to true single photon sources), loss in the transmission line and non-ideal detectors [29]. Most commercial quantum links have two detectors, each tuned to detect protons in one of the two different polarisation states - "1" or "0" - used to make up the secret code. Hoi-Kwong Lo at the University of Toronto in Canada realized [4] that small imperfections in the design of the photon detectors mean they aren't quite switched on at the same instant, and for a few picoseconds only one will be on. Eve can make sure the photon arrives at Bob when only his "1" detector is open. Now, if Bob registers a click and tells Alice, Eve knows that the photon was in the "1" state. Lo claims that their team was able to hack a commercial quantum communications device 4% of the time. Just recently id Quantique said that they were able to fix loopholes exploited by Professor Lo.[21].

Eve is assumed to know everything about Alice's and Bob's equipment. Thus, Eve can fully exploit every imperfection that exists in legitimate parties' hardware and software.

Makarov [11],[12] explores successful attacks on commercial single-photon de-tectors, using fake state pulses, utilizing equipment imperfections. Using bright light

Eve can blind Bob's detectors forcing them to become totally insensitive to single photons as well as dark counts and afterpulses, only producing an output pulse (a "click") when a brighter optical pulse is applied at its input. With such a control mode Eve could intercept each quantum bit encoded by Alice with an exact replica of the detection apparatus used by the Bob, then send a faked state targeting the corresponding detector at the receiver's side, allowing Eve to get a complete copy of the cryptographic key without being noticed unless light intensity across the link is monitored.

Finally, QKD security is always relying on an implicit assumption: Alice and Bob, who are storing the final symmetric secret keys in classical memories must be located inside secure environments. If there is a channel allowing to spy on the keys, stored in a classical memory, then the security of the keys is compromised. Providing that QKD devices are partly made of classical objects, it is essential that such interfaces are designed with great care.

22.6 Other Challenges

Speed of key exchange and reachable distance of QKD links are challenging factors today. According to SECOQC reports as of 2007 [13] one can expect to exchange between 1 and 10 kbits of secret key per second, over a point-to-point QKD link of 25 km (at 1550 nm, on dark fibres). The maximum span of QKD links is roughly 100 km at 1550 nm on telecom dark fibres. This range is suitable for metropolitan area scale QKD. Both secret bit rate and maximum reachable distance are expected to continue their progression during the next years due to combined theoretical and experimental advances. Significant speed increase is expected in forthcoming future, though it will require very fast detectors at telecommunications wavelengths, with good quantum efficiency and low dark count.

Use of trusted relays QKD network can increase distance reacheable by QKD link. [13]. The relay nodes need to be trusted, although trust can be reduced by having the sender use a secret sharing scheme. It is particularly useful when the network operator is already a network user, as in the case of internal bank networks. Global key distribution is performed over a QKD path, i.e. a one-dimensional chain of trusted relays connected by QKD links, establishing a connection between two end nodes, as shown on Fig. 22.18. Secret keys are forwarded, in a hop-by-hop fashion, along QKD paths. To ensure their secrecy one can use one-time pad encryption

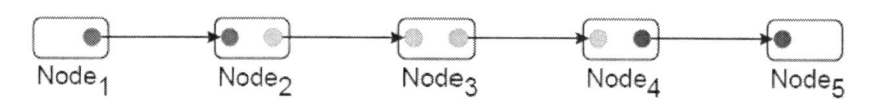

Fig. 22.18 "Hop-by-hop" secure message passing by QKD links

and unconditionally secure authentication, both realised with a local QKD key. The trusted relays QKD network has been used in the DARPA and Vienna Network.

One of the main targets of the free-space QKD system is to construct an Earth-satellite link. Several groups have published detailed modeling to show that low-Earth orbit satellite-to-ground QKD would be feasible even in daylight, with typical ranges of 1,000 km [15].

Another challenge can be posed by Key Pre-Distribution for system initialization. After that QKD-generated keys can then be stored and used for later authentication. For the network of **n** nodes this may lead to **n(n-1)/2** pairs of secret keys distributed, but thanks to possibility of playing with betwork connectivity, the problem can be reduced to linear one. [13].

22.7 Conclusions

Quantum key distribution solves the key distribution problem with security based on the laws of physics, but it is important to develop network architecture able to fully benefit from the possibilities offered by point-to-point, distance limited QKD links.

Few experimental demonstrations have included all of the ingredients of a full QKD protocol, and their focus has been almost exclusively on closing the gap between the idealized assumptions of "theoretical secrecy" proofs for QKD and the realities of imperfect realizations of fundamental quantum processes. As the technology continues to evolve into more mature stage, it is apparent that QKD is capable of significantly and positively impacting information-security requirements without insisting on theoretically perfect secrecy from inevitably imperfect physical realizations.

According to a roadmap projected by the Quantum Cryptography Technology Experts Panel [14], at least two distinct practical roles for QKD are possible within future networked optical communications infrastructures "key-transfer-mode QKD": an enhancement to conventional key management infrastructures supporting the transfer or generation of keys for symmetric key cryptography "encryptor-mode QKD": a new, physical layer encryption technology (a quantum generated one-time-pad stream cipher).

The roadmap sets out specific goals that will stimulate the necessary basic theoretical and experimental physics research and advances in the enabling component technologies. The roadmap has been a living document, updated on an annual basis to reflect progress.

The latest Updating Quantum Cryptography Working Group Report [30] outlines the standardization of quantum cryptography. Specifically it raises issues of the interoperability specifications and requirements. One is the interoperability between quantum cryptographic technology and contemporary cryptographic systems and the other is that among quantum cryptosystems. It also refers to issues relating to test requirements.

Research goals related to New Generation Quantum Cryptography define a short-term strategy that is to combine current quantum cryptography and photonic network technology with reasonable assumptions for the nodes and a compromise of the security level. The long-term strategy is to invent new schemes that have the merits of all known protocols and study and develop quantum repeaters that can realize full quantum networking.

References

1. Recommendation for Key Management - Part 1: general, NIST Special Publication 800-57 (March 2007),
   http://csrc.nist.gov/publications/nistpubs/800-57/ SP800-57-Part1.pdf
2. Shor, P.W.: Polynomial-Time Algorithms for Prime Factorization and Discrete Logarithms on a Quantum Compute. SIAM J. Computing 26 (1997)
3. Elliott, C., Pearson, D., Troxel, G.: Quantum Cryptography in Practice. BBN (2003)
4. Vakhitov, A., Makarov, V., Hjelme, D.R.: Large pulse attack as a method of conventional optical eavesdropping in quantum cryptography. Journal of Modern Optics 48, 2023–2038 (2001)
5. Paterson, K.G., Piper, F., Schack, R.: Why Quantum Cryptography? Cryptology ePrint Archive: Report 2004/156, http://eprint.iacr.org/2004/156
6. Id Quantique: Cerberis (2008),
   http://www.idquantique.com/products/cerberis.htm
7. MagicQ, Inc.: MagiQ QPN Security Gateway (2007), http://www.magiqtech.com
8. Houston, L.: Secure Ballots Using QC (2007), http://www.cse.wustl.edu/~jain/
9. Lomonaco, S.J.: A quick glance at quantum cryptography. Univ. of Maryland (November 1998)
10. Bennett, C.H.: Quantum Cryptography: Uncertainty in the Service of Privacy. Science 257, 752–753 (1992)
11. Sauge, S., Makarov, V., Anisimov, A.: Quantum hacking: how Eve can exploit component imperfections to control yet another of Bob's single-photon qubit detectors. In: CLEO Europe EQEC (2009)
12. Makarov, V.: Controlling passively-quenched single photon detectors by bright light. arXiv: 0707.3987v3 [quant-ph] (April 2009)
13. SECOQC White Paper on Quantum Key Distribution and Cryptography. Secoqc-WP-v5 (January 2007)
14. A Quantum Information Science and Technology Roadmap Part 2: Quantum Cryptography, Report of the Quantum Cryptography Technology Experts Panel, ARDA (July 2004)
15. Rarity, J.G., Tapster, P.R., Gorman, P.M., Knight, P.: Ground to satellite secure key exchange using quantum cryptography. New Journal of Physics 4, 82.1–82.9 (2002)
16. Bennett, C.H., Brassard, G.: Quantum cryptography: Public key distribution and coin tossing. In: International Conference on Computers, Systems & Signal Processing, Bagalore, India, December 10-12, pp. 175–179 (1984)
17. Inoue, K., Waks, E., Yamamoto, Y.: Differential Phase Shift Quantum Key Distribution. Physical Review Letters 89, 37–902 (2002)
18. Scarani, V., Acin, A., Ribordy, G., Gisin, N.: Quantum Cryptography Protocols Robust against Photon Number Splitting Attacks for Weak Laser Pulse Implementations. Physical Review Letters 92, 57–901 (2004)
19. Hwang, W.-Y.: Quantum key distribution with high loss: Toward global secure communication. Phys. Rev. Lett. 91(5), 057901 (2003)

20. Wang, X.-B.: Beating the Photon-Number-Splitting Attack in Practical Quantum Cryptograpy. Physical Review Letters 94, 230–503 (2005)
21. Barras, C.: Quantum computers get commercial - and hackable, New Scientist (April 2009)
22. Townsend, P.D., Thompson, I.: A quantum key distribution channel based on optical fibre. Journal of Modern Optics 41(12), 2425–2433 (1994)
23. Jacobs, B.C., Franson, J.D.: Quantum cryptography in free space. Optics Letters 21, 1854–1856 (1996)
24. Bennett, C.H.: Quantum cryptography using any two nonorthogonal states. Physical Review Letters 68(21-25), 3121–3124 (1992)
25. Ekert, A.K.: Quantum cryptography based on Bell's theorem. Physical Review Letters 67(6), 661–663 (1991)
26. Gisin, N., Ribordy, G., Tittel, W., Zbinden, H.: Quantum cryptography. Review of Modern Physics 74 (2002)
27. Anscombe, N.: Quantum cryptography: Vienna encrypts communication network. OLE (January 2009), http://www.optics.org/ole
28. Zhao, Y., Bing, Q., Xiongfeng, M., Hoi-Kwong, L., Li, Q.: Experimental Decoy State Quantum Key Distribution Over 15km. arXiv: 0503.192v2 [quant-ph] (March 2005)
29. Makarov, V., Hjelme, D.: Faked states on quantum cryptosystems. J. Mod. Opt. 45, 2039–2047 (2001)
30. The Updating Quantum Cryptography Report. Ver. 1 (May 2009)

Chapter 23
Cerberis: High-Speed Encryption with Quantum Cryptography

Leonard Widmer

Abstract. Cerberis is a scalable solution that combines high speed layer 2 encryption appliance with the power of quantum key distribution (QKD) technology. Additional encryption appliances can be added to a QKD server at any time, without network interruption. It allows company for a scalable deployment when more bandwidth or additional protocols is requested. Cerberis is a solution that offers a radically new approach to network security based on a fundamental principle of quantum physic - i.e. observation causes perturbation to provide unprecedented security.

23.1 Cryptography

With the development of electronic and optical telecommunication networks, the quantity of information exchanged and the reliance of organizations on these new communication channels has increased exponentially. Simultaneously, the risk associated with espionage of sensitive information grew significantly. As a result, technologies were developed to cope with these threats.

The need for high speed links is increasing and thus optical fiber cables have replaced copper cables. Whether they are aware of it or not, organizations almost certainly rely on optical fibers to transmit their data.

Sensitive data travels across the network of multiple telecom companies before they reach their final destination. Optical fiber cables are thus often used to carry confidential information. These links are still frequently left unprotected, with confidential data being sent in clear, because of the lack of adequate security solutions and a false perception on the difficulty of tapping such links.

Intercepting information transmitted over an optical fiber cable is not only possible but also quite easy in practice, without even interrupting the line, by using cheap devices also known as optical tap, available on the market. As optical fiber

Leonard Widmer
id Quantique SA, ch. de la Marbrerie 3, 1227 Carouge, Switzerland
e-mail: leonard.widmer@idquantique.com

E. Tkacz and A. Kapczynski (Eds.): Internet – Technical Development and Appl., AISC 64, pp. 217–221.

Fig. 23.1 Optical tap to intercept data on optical fiber

links constitute a dangerous loophole in the information security infrastructure of an organization, CxOs must ensure that mission critical data are protected when travelling outside the secure perimeter of the company by using the strongest encryption technique available, without impacting network performance.

23.2 Organizational Information-Transition Nets

Cryptography is the art of rendering information exchanged between two parties unintelligible to any unauthorized person. Although confidentiality is the traditional application of cryptography, it is used nowadays to achieve broader objectives, such as authentication, digital signatures and no repudiation.

The way cryptography works is illustrated below. Before transmitting sensitive information, the sender combines the plain text with a secret key, using some encryption algorithm, to obtain the cipher text. This scrambled message is then sent to the recipient who reverses the process to recover the plain text by combining the cipher text with the secret key using the decryption algorithm. An eavesdropper cannot deduce the plain message from the scrambled one without knowing the key. As one usually assumes that the encryption algorithm is disclosed, the secrecy of such a scheme basically depends on the fact that the key is secret. This means first,

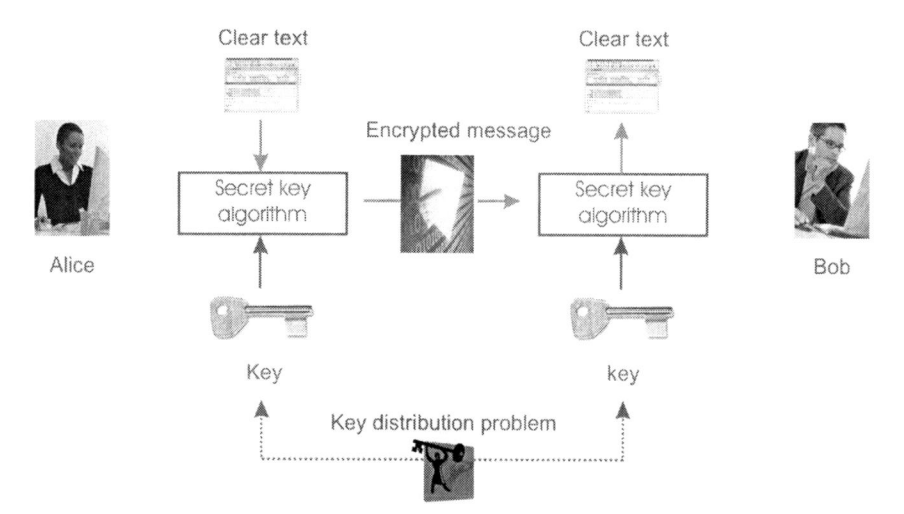

Fig. 23.2 Principle of cryptography

that the key generation process must be appropriate, in the sense that it must not be possible for a third party to guess or deduce it. Truly random numbers must thus be used for the key.

Second, it must not be possible for a third party to intercept the key during its exchange between the sender and the recipient. This so-called "key distribution problem" is very central in cryptography.

23.3 Quantum Cryptography

The Achilles heel of existing cryptographic solutions is the key exchange process. While conventional key distribution techniques rely on the difficulty of certain mathematical problems, and therefore offer only conditional security, the secrecy of keys distributed by quantum cryptography is guaranteed by the fundamental laws of quantum physics. According to the Heisenberg Uncertainty Principle, one of the central principles of quantum physics, it is not possible to observe a quantum object without modifying it.

Quantum cryptography is a technology that exploits this fundamental principle of quantum physics - observation causes perturbation - to ensure future-proof key exchange for symmetric cryptography algorithms (e.g. AES) on a point to point communication over optical fibre link.

Quantum cryptography solves the key distribution problem; therefore quantum key distribution (QKD) is a more appropriate terminology.

23.4 Advantage of Layer 2 Encryption

There are several ways to encrypt data in motion; options include Secure Sockets Layer (SSL) for the Internet and the IPSec standard (or layer 3 encryption) for "tunnelling" and layer 2 encryption.

The word "layer" refers to the way in which network communication systems are designed, for example layer 1 is the physical layer (wire, cables, connectors etc), layer 2 is the data link layer (eg Ethernet frames) & layer 3 is the network layer (eg IP packets).

In reality, encryption can happen at different layers of a network stack, the following are just a few examples:

- End-to-end encryption happens within applications.
- SSL encryption takes place at the transport layer.
- IPSec encryption takes place at the network layer.
- Layer 2 encryption takes place at the data link layer.

The challenge lies in maintaining the performance and simplicity of high speed networks while assuring the security and privacy of user data, whether it is a voice, data or video transmission.

Layer 2 encryption is often referred to as a "bump in the wire" technology. The phrase conveys the simplicity, maintainability and performance benefits of layer 2 solutions that are designed to be transparent to end users with little or no performance impact on network throughput. It is proven that Layer 2 encryption technologies provide superior throughput and far lower latency than IPSec VPNs, which operate at Layer 3.

Outsourcing of routing tables is also seen as a weakness of Layer 3 VPN services because many corporations don't want to relinquish control or even share their routing schemes with anyone, not even their service provider. They prefer Layer 2 network services, such as Ethernet MPLS, frame relay or ATM as these are simpler in architecture and that allow customers to retain control of their own routing tables.

23.5 Description of the Cerberis Solution

Cerberis is a fast and secure solution that combines high-speed layer 2 encryption with quantum key distribution (QKD).

The Cerberis solution is cost-effective as it evolves with the network. Additional encryption appliances can be added to a QKD server at any time, without network interruption. This allows for a scalable deployment, adding more encryption appliances whenever necessary to increase the bandwidth or to add additional protocols, without upgrading the QKD server.

With the Cerberis solution, your infrastructure investments last longer and your total cost of ownership is reduced.

Fig. 23.3 Cerberis; QKD server with high-speed encryption appliance

23.5.1 Quantum Key Distribution Server (QKD Server)

The exchange of secret encryption keys is performed in an appliance, called the QKD server. A fundamental principle of quantum physics - observation causes perturbation - is exploited to exchange secret keys between two remote parties over an optical fibre with unprecedented security. The QKD server autonomously produces, manages and distributes secret keys to one or more encryption appliances through a secure dedicated channel.

QKD is a point to point technology and the two servers must be linked by a dark fibre with maximum length of approximately 80km.

23.5.2 High Speed Encryption Engine (AES-256)

A 2nd dedicated appliance, the Centauris, perform high-speed encryption based on the standardized Advanced Encryption Standard (AES). Point-to-point wire-speed encryption with low latency and no packet expansion is made possible by operating at the layer 2 of the OSI model. Four protocols are supported, namely Gigabit Ethernet (up to 10Gbps), Fibre channel (up to 4 Gbps), SONET/SDH (up to 10Gbps) and ATM (up to 622Mbps).

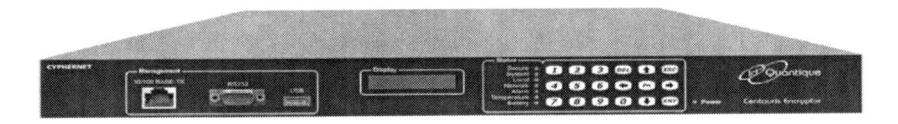

Fig. 23.4 High-speed encryption appliance

23.6 Conclusions

Cerberis combines strong high-speed encryption with quantum key distribution and makes it possible to achieve an unprecedented level of security. These techniques are particularly applicable for high value applications and long term secure data retention requirements.

References

1. Marcikic, I., Lamas-Linares, A., Kurtsiefer, C.: Free-Space Quantum Key Distribution with Entangled Photons. Appl. Phys. Lett. 89, 101122 (2006)
2. Rarity, J.G., Tapster, P.R., Gorman, P.M., Knight, P.: Ground to Satellite Secure Key Exchange using Quantum Cryptography. New J. Phys. 4, 82 (2002)
3. Ursin, R., Tiefenbacher, F., Schmitt-Manderbach, T., Weier, H., Scheidl, T., Lindenthal, M., Blauensteiner, B., Jennewein, T., Perdigues, J., Trojek, P., Omer, B., Furst, M., Meyenburg, M., Rarity, J., Sodnik, Z., Barbieri, C., Weinfurter, H., Zeilinger, A.: Entanglement-Based Quantum Communication Over 144km. Nat. Phys. 3, 481–486 (2007)

Chapter 24
A Numerical Simulation of Quantum Factorization Success Probability

Piotr Zawadzki

Abstract. The quantum factorization is probably the most famous algorithm in quantum computation. The algorithm succeeds only when some random number with an even order relative to factorized composite integer is fed as the input to the quantum order finding algorithm. Moreover, post processing of the quantum measurement recovers the correct order only for some subset of possible values. It is well known that numbers with even orders are found with probability not less than $1/2$. However, numerical simulation proves that probability of such event exhibits grouping on some discrete levels above that limit. Thus, one may conclude that usage of the lowest estimate leads to underestimation of the successful factorization probability. The understanding of the observed grouping requires further research in that field.

24.1 Introduction

Theoretical study of quantum systems serving as computational devices has achieved tremendous progress in the last few years. We now have strong theoretical evidence that quantum computers might be used as a powerful computational tool, capable of performing tasks which seem intractable for classical computers. There are two well known quantum algorithms that can be performed more efficiently than their classical counterparts, namely Shor's order finding [1] and Grover's database search [2]. The order finding seems to be the most stimulating development in the field of quantum information processing as it provides exponential to polynomial complexity reduction of the problem. Polynomial time factorization with the aid of both a quantum computer and a classical one is a direct consequence of efficient

Piotr Zawadzki
Silesian University of Technology, Institute of Electronics,
Akademicka 16, 44-100 Gliwice, Poland
e-mail: Piotr.Zawadzki@polsl.pl

E. Tkacz and A. Kapczynski (Eds.): Internet – Technical Development and Appl., AISC 64, pp. 223–231.

order finding. Interest in the factoring problem is especially great for composite integers being a product of two large prime numbers – the ability to factor such integers is equivalent to the ability to read information encoded via the RSA cryptography system [3]. Thus, quantum computers, if built, pose a serious challenge for the security of today's asymmetric cryptographic systems. The quantum order finding is a probabilistic process because of the inherent nature of the quantum measurement. The researchers in the field are mainly interested in the success probability of the order recovery and proposed many modifications improving that aspect of the original Shor's proposal [4, 5]. It is reported that probability of the order recovery may be extremely high if multiple quantum measurements and sophisticated post processing are employed [6]. However, the measurement of the quantum system is not the only source of the probabilistic behavior of the algorithm. The factorization is possible only if some random number fed as the input to the algorithm has an even order. The lower bound on finding parameter suitable for factorization is given in [7]. The aim of this paper is to provide an analysis of the randomness introduced by that classical part of the algorithm. Multiple quantum measurements require repetitive runs of a quantum computer which is undoubtedly costly in terms of money and effort. The realistic estimate of algorithm success should assume only one run of a quantum device. The Sect. 24.3 provides numerical simulations of Shor's algorithm success probability under such assumption.

24.2 Quantum Factorization

The factorization reduction to order finding exploits the following observation. Let N be the composite number and $x < N$ is coprime to N. The order of x is the smallest number such that $x^r \bmod N = 1$. If the order of x is even then one may write

$$\left(x^{r/2} - 1\right)\left(x^{r/2} + 1\right) \bmod N = 0 \tag{24.1}$$

and

$$p = \gcd\left(x^{r/2} - 1, N\right) \tag{24.2}$$

is a nontrivial factor of N provided that

$$x^{r/2} \bmod N \neq -1 \ . \tag{24.3}$$

Thus factorization of composite number N is reduced to the order finding of some number x. Classical order finding gives no advantage over other factorization algorithms as its complexity is also exponential. However, it is possible to determine the order of x in polynomial time by the quantum phase estimation algorithm applied to modular multiplication function followed by order recovery with the continued fraction expansion algorithm.

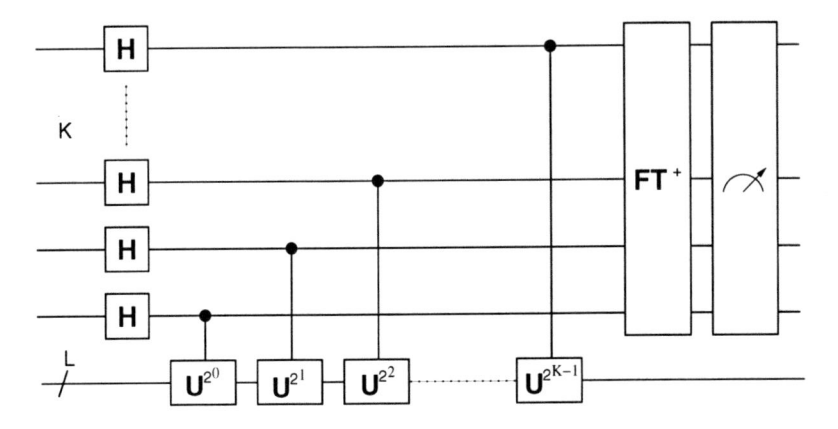

Fig. 24.1 Quantum circuit for phase estimation

24.2.1 Phase Estimation

The problem statement is as follows. Given the unitary operator $\mathbf{U}$ and its eigenvector $|u\rangle$ find the phase φ of the eigenvalue

$$\mathbf{U}|u\rangle = e^{2\pi j\varphi}|u\rangle \ . \tag{24.4}$$

The solution is based on the observation that controlled-$\mathbf{U}$ operation extracts unknown eigenvalue to the control register if the target register is initialized with the eigenvector

$$\mathbf{CU}\left(|0\rangle + |1\rangle\right) \otimes |u\rangle = \left(|0\rangle + e^{2\pi j\varphi}|1\rangle\right) \otimes |u\rangle \ . \tag{24.5}$$

Assume that one can efficiently implement application of $\mathbf{U}$ operator p times for p being power of 2 and consider the circuit from Fig. 24.1. Let the control register of size K be initialized to the state $|0\rangle^{\otimes K}$ and target register to the eigenstate of the operator $\mathbf{U}$. The application of Hadamard gates and series of controlled $\mathbf{U}^p$ operators results in the following control register

$$\frac{1}{\sqrt{M}}\left(|0\rangle + e^{2\pi j2^{K-1}\varphi}|1\rangle\right) \otimes \ldots \otimes \left(|0\rangle + e^{2\pi j2^{0}\varphi}|1\rangle\right) = \frac{1}{\sqrt{M}}\sum_{k=0}^{M-1} e^{2\pi jk\varphi}|k\rangle \tag{24.6}$$

where $M = 2^K$ and $k = k_{K-1}2^{K-1} + \ldots + k_0 2^0$. The application of quantum inverse Fourier transform gives

$$\mathbf{FT}^{+}\left(\frac{1}{\sqrt{M}}\sum_{k=0}^{M-1} e^{2\pi jk\varphi}|k\rangle\right) = \frac{1}{\sqrt{M}}\sum_{k=0}^{M-1} e^{2\pi jk\varphi}\frac{1}{\sqrt{M}}\sum_{l=0}^{M-1} e^{-2\pi j\frac{lk}{M}}|l\rangle =$$

$$= \frac{1}{M}\sum_{l=0}^{M-1}\left(\sum_{k=0}^{M-1} e^{-2\pi j(\frac{l}{M}-\varphi)k}\right)|l\rangle =$$

$$= \frac{1}{\sqrt{M}} \sum_{l=0}^{M-1} \left(\frac{\sin\left[\pi\left(\frac{l}{M} - \varphi\right)M\right]}{\sin\left[\pi\left(\frac{l}{M} - \varphi\right)\right]} \right) e^{\pi j \gamma} |l\rangle \quad (24.7)$$

where $\gamma = (l/M - \varphi)(M-1)$. Probability of measuring the state $|l\rangle$

$$p(l) = \frac{1}{M^2} \left[\frac{\sin\left[\pi\left(\frac{l}{M} - \varphi\right)M\right]}{\sin\left[\pi\left(\frac{l}{M} - \varphi\right)\right]} \right]^2 \quad (24.8)$$

has sharp maximum when $l/M \approx \varphi$. In fact, if the unknown phase φ may be represented on K bits then exist l providing exact estimation with probability 1. Otherwise, the estimation has small but finite probability of failure. It is proved that phase estimation with accuracy up to L bits with failure probability less then ε requires the control register of size [8]

$$K = L + \log_2\left[2 + 1/(2\varepsilon)\right] \quad . \quad (24.9)$$

If the target register is the superposition of operator $\mathbf{U}$ eigenstates then the above procedure leads to the phase estimation of some randomly selected eigenvalue.

24.2.2 Quantum Order Finding

Define the operator $\mathbf{U}$ as

$$\mathbf{U}|y\rangle = |xy \bmod N\rangle \quad . \quad (24.10)$$

If $|y\rangle = |1\rangle$ then the left hand side of the circuit from Fig. 24.1 performs modular exponentiation for all possible powers at once

$$\mathbf{CU}^{2^{K-1}} \ldots \mathbf{CU}^{2^{K-1}} \left(\mathbf{H}^{\otimes K} |0\rangle \right) |1\rangle =$$

$$= \mathbf{CU}^{2^{K-1}} \ldots \mathbf{CU}^{2^{K-1}} \left(\frac{1}{\sqrt{M}} \sum_{l_0,\ldots,l_{K-1}=\{0,1\}} |l_{K-1}\rangle \ldots |l_0\rangle \right) |1\rangle =$$

$$= \frac{1}{\sqrt{M}} \sum_{l_0,\ldots,l_{K-1}=\{0,1\}} |l_{K-1}\rangle \ldots |l_0\rangle |x^{l_{K-1} 2^{K-1}} \ldots x^{l_0 2^0} \bmod N\rangle =$$

$$= \frac{1}{\sqrt{M}} \sum_{l=0}^{M-1} |x^l \bmod N\rangle |l\rangle \quad . \quad (24.11)$$

The right hand side of the above equation is periodic with period equal to order of x relative to composite number N. The eigenstates $|u_s\rangle$ of operator $\mathbf{U}$ satisfy [9]

$$\mathbf{U}|u_s\rangle = e^{2\pi j \varphi_s} |u_s\rangle \quad ,$$

$$\frac{1}{\sqrt{r}} \sum_{s=0}^{r-1} |u_s\rangle = |1\rangle \quad (24.12)$$

where $\varphi_s = s/r$ and r is the order of x. The phase estimation circuit with target register set to $|u_s\rangle$ returns the estimate l/M of the φ_s (s determined by $|u_s\rangle$ and r determined by x). The algorithm of continued fraction expansion applied to estimate l/M recovers phase s/r only when s and r are coprime and

$$\left| \frac{l}{M} - \frac{s}{r} \right| \leq \frac{1}{2r^2} \, . \tag{24.13}$$

It follows from the above that accuracy of L bits in order determination requires of $2L+1$ bits in the base estimation procedure. Taking into account success probability of the phase measurement (24.9), one concludes that order finding with precision of L bits and failure probability less than ε requires control register of size

$$K = 2L + 1 + \log_2 \left[2 + 1/(2\varepsilon) \right] \, . \tag{24.14}$$

However, it is not known how to construct the eigenstate $|u_s\rangle$. Moreover, modular exponentiation requires initialization of the target register with state $|1\rangle$ – the equiprobable superposition of operator $\mathbf{U}$ eigenstates (see equation (24.12)). Thus phase estimation procedure returns eigenvalue phase estimate for some randomly selected eigenstate, and in consequence, a random number s.

24.2.3 The Shor's Algorithm

The following steps summarize the Shor's algorithm for quantum factorization of the composite number N:

1. Select a random number x coprime to N (otherwise $\gcd(x,N)$ is a factor of N). Only some x are good candidates as the order of x determined by the next step has to be even and condition (24.3) must be fulfilled.
2. Find the order of x with the quantum computer. The order is successfully recovered only for some subset of valid quantum measurements.
3. Calculate divisor p from the equation (24.2) and return to the point 1) with $N = N/p$.

It is clear, that the nature of the above algorithm is probabilistic even if one assumes infinite accuracy of the quantum measurement. The success rate of the algorithm depends on the following classical random factors:

- the selection of the "lucky" x with an even order which fulfils condition (24.3),
- the success of the order recovery from the phase estimate with the continued fraction expansion.

The lower bound on success probability of the first step is given in [7] as

$$p(x) \geq 1 - \frac{1}{2^{k-1}} \tag{24.15}$$

where k is the number of prime factors of N. That lower bound has maximal value when composite number is a product of only two prime numbers, what in fact represents the most interesting situation. Sect. 24.3 provides more information about that probability distribution. The possible outcomes probability distribution exhibits sharp peaks in the vicinity of the phases to be estimated. However, there exists nonzero probability that the measurement will fail. The failure probability is directly related to the width of peaks that in turn depends on the control register size [9]. Thus the confidence of quantum measurement may be enlarged to arbitrary accuracy by appropriate selection of the control register size. The success of the phase estimation also depends on the order value. Assuming an infinite accuracy of the quantum measurement and single use of quantum device, the continued fraction expansion algorithm provides correct order recovery from the phase estimate l/M only when the convergent $\varphi = s/r$ is formed by coprime numbers. Otherwise, one obtains order underestimated by a factor equal to greatest common divisor of s and r. The quantum measurement selects one of the states of the target register by random. The number of s coprime to r is given by Euler's totient function $\Phi(r)$. In consequence, the order of x may be recovered from phase estimate with probability

$$p(r) = \Phi(r)/r \ . \tag{24.16}$$

The probability of the successful factorization in the first run of the algorithm is given by

$$p(N) = (1 - \varepsilon) \sum_{\text{even } r} p(r)p(x|r) \tag{24.17}$$

where $p(x|r)$ is the selection probability of x with order r and ε is the quantum measurement failure probability.

24.3 Analysis

It is interesting to estimate the factorization success probability for composite numbers of the form $N = pq$, where p and q are primes. Such numbers play a key role in many asymmetric cryptography algorithms. The provided analysis was performed for products of p and q taken from the list of the first 100 prime numbers. The key factors determining the factorization success are:

- probability $p(r)$ that phase estimation for the given r is successful,
- probability $p(x|r)$ of finding $x < N$ with even order r and satisfying (24.3).

The Fig. 24.2 presents $p(r)$ calculation results for even orders. It is worth noting that probability calculated from (24.16) never exceeds the value of $1/2$. Such behavior comes from the fact, that only even orders are meaningful for factorization purposes, and in the totient function representation resulting from fundamental theorem of arithmetic

$$\Phi(r) = r \sum_{k} (1 - 1/p_k) \tag{24.18}$$

at least one of prime factors p_k is equal to 2 for even orders.

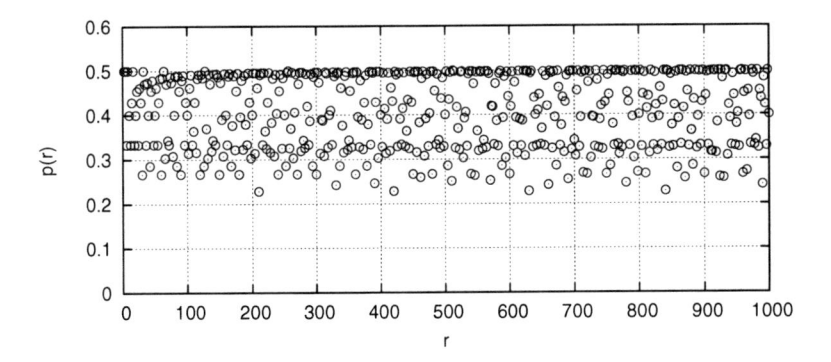

Fig. 24.2 Phase estimation success probability

It follows from (24.15) that for composite odd number of the form $N = pq$ and x coprime to N probability of finding x suitable for factorization is greater than $1/2$. However, simulation results presented on Fig. 24.3 exhibit a deeper structure in that probability distribution. The grouping of points around some discrete levels is evident and that suggests the existence of a class of composites less resistant to quantum factorization than others. Observed behavior probably comes from the features of numbers $(p-1)$ and $(q-1)$, however, there is no satisfactory explanation of Fig. 24.3 in the literature known to the author. The profound number theory study of the order finding procedure is thus required.

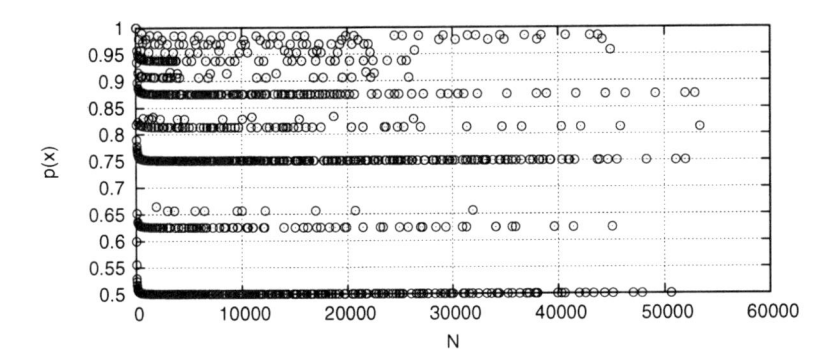

Fig. 24.3 Probability of finding x with even order

The Fig. 24.4 presents the success rate of the Shor's factorization algorithm calculated from (24.17) with an assumption of the 100% certainty of the quantum measurement and single usage of the quantum device.

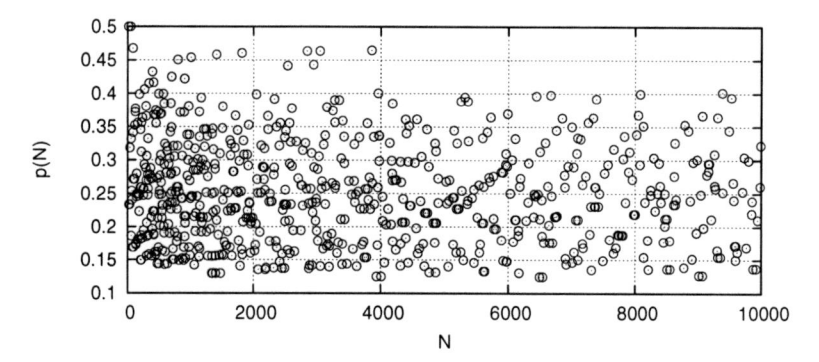

Fig. 24.4 The factorization success probability in the first run of Shor's algorithm

24.4 Conclusion

The quantum factorization algorithm represents a breakthrough in complexity theory and modern cryptography. The Shor's algorithm owes its fame to polynomial time breaking of virtually all presently used public key algorithms. Unfortunately, the practical breaking is out the reach by now because factorization of number 15 is still one of the most complicated quantum computation [10]. However, very rapid progress in that field is observed so it is difficult to estimate the time horizon when practical computation will be in scientists' reach. The quantum factorization was analyzed many times and there were proposed modifications to the original algorithm version improving its speed and efficiency. However, the researchers were concentrated so far on the probabilistic aspect of the quantum measurement. The randomness introduced by the classical parts of the algorithm still requires further investigation. The computer simulation results presented herein expose that success rate of the algorithm is usually underestimated.

References

1. Shor, P.W.: Polynomial-time algorithms for prime factorization and discrete logarithms on a quantum computer. SIAM J. Sci. Comput. 26, 1484–1509 (1997)
2. Grover, L.K.: A fast quantum mechanical algorithm for database search. In: 28th Annual ACM Symposium on the Theory of Computing, pp. 212–219 (1996)
3. Gerjuoy, E.: Shor's factoring algorithm and modern cryptography. An illustration of the capabilities inherent in quantum computers. A. J. Phys. 73, 521–540 (2005)
4. Knill, E.: On Shor's quantum factor finding algorithm: Increasing the probability of success and tradeoffs involving the Fourier Transform modulus, Technical Report LAUR-95-3350, Los Alamos Laboratory (1995),
   http://www.eskimo.com/~knill/cv/reprints/knill:qc1995c.ps
5. McAnally, D.: A refinement of Shor's algorithm (2001),
   http://xxx.lanl.gov/pdf/quant-ph/0112055

6. Bourdon, P.S., Williams, H.T.: Probability estimates for Shor's algorithm. Quant. Inf. Comput. 7, 522–550 (2007)
7. Ekert, A., Jozsa, R.: Quantum computation and Shor's factoring algorithm. Rev. Mod. Phys. 68, 733–753 (1996)
8. Nielsen, M.A., Chuang, I.L.: Quantum Computation and Quantum Information. Cambridge University Press, Cambridge (2000)
9. Desurvire, E.: Classical and Quantum Information Theory. Cambridge University Press, Cambridge (2009)
10. Vandersypen, L.M.K., Steffen, M., Breyta, G., Yannoni, C.S., Sherwood, M.H., Chuang, I.L.: Experimental realization of Shor's quantum factoring algorithm using nuclear magnetic resonance. Nature 414, 883–887 (2001)

Chapter 25
Software Flaws as the Problem of Network Security

Teresa Mendyk-Krajewska and Zygmunt Mazur

Abstract. The software flaws give the possibility of attacking the IT systems, which may result in gaining remote control of them. The known vulnerabilities may be detected with the use of scanners and the threat is then removed through installing some patch or a new software version; however, when the tool does not have the updated threat database, the effect of its functioning may give illusory security. Furthermore, in practice, new vulnerabilities in the security systems are identified all the time, which indicates that obtaining high level of security is a significant problem.

25.1 The Vulnerability of Systems to Hacking

The reasons of the vulnerability of computer systems to hacking include the flaws of software, mainly of the operating systems, their components and the network protocols. The software systems are increasingly complex and the vulnerabilities result from both the errors made by the programmers and the insufficient testing of ready products, which is, in turn, the effect of strong competition among companies (which means the need of quick promotion of new software). The problem may be also caused by the desire to reduce the costs. The research reveals that only every third organization preserves the correct software lifecycle in the development phase, which allows for increasing the security of applications due to running the appropriate number of tests prior to their implementation. The faulty configuration of the installed software also creates the vulnerabilities, which leads to faulty operation of the system or the decrease of its stability. The software flaws allow for unauthorized remote access to the system with the use of specially prepared code (the co-called exploit). An exploit allows for bypassing the security mechanisms and, in consequence, for gaining control of the system. These flaws in software are characterized

Teresa Mendyk-Krajewska · Zygmunt Mazur
Wroclaw University of Technology,
Faculty of Computer Science and Management, Institute of Informatics
e-mail: {Teresa.Mendyk-Krajewska,Zygmunt.Mazur}@pwr.wroc.pl
http://www.ii.pwr.wroc.pl

E. Tkacz and A. Kapczynski (Eds.): Internet – Technical Development and Appl., AISC 64, pp. 233–241.

with some cycle which includes distinguishable stages. After the discovery of a flaw allowing for carrying out the attack on the system, the first exploits appear and they are used by the hackers who have just started hacking. Then the automatic tools are created which use a given vulnerability and after that they are extensively used. Up to this time, the number of successful attacks is proportionally increasing. After the appearance of the first exploits, the security patch is released; the time from the discovery of the vulnerability to its fixing with a patch may differ. The research reveals that long after the patch release, the number of attacks carried out with the use of a given vulnerability increases [1].

In March 2009 in Vancouver, the CanSecWest conference was held and its programme included a contest in the scope of breaking into the computers with the most popular operating systems. As early as on the first day it was proved that any web browser may be used to gain control of the system. For instance, the record holders broke into with the use of Safari in about two minutes. The operating systems were also tested during the conference and it turned out that the most vulnerable one is Mac OS [3].

The report for 2008 which was prepared by the National Infrastructure Protection Center, the FBI agency dealing with the network security, reveals that the hackers usually do not apply any complex methods but they make use of known vulnerabilities in the computer systems which are not fixed with available patches and of popular tools. Meanwhile, the information about the discovery of other flaws in the systems is announced all the time. The research carried out in April 2009 by Forrester Research shows that during the last year over 62% of companies experienced the violation of IT system security due to the vulnerabilities in the software. Also the results of the research conducted in the United States and Great Britain and ordered by Veracode company, which provides the application risk management systems, (180 organizations operating in different branches of industry were questioned) indicate that the vulnerabilities in applications are the main cause of violating the security [4].

25.2 The Attacks on the Operating Systems

One of the most frequent goals of attacks includes the operating systems. Various types of attacks have slightly different course; however, it is possible to outline some scenario of the attack on an operating system. It usually includes the following steps:

- conducting the reconnaissance (locating and gathering the information about the system),
- making use of errors and vulnerabilities in the system of the software (the operating system, components or application) or its configuration,
- breaking into the account of a legal user,
- increasing the authorizations, if necessary,
- performing the unauthorized actions,

- installing the gate for on-going or future use,
- deleting the traces of activity (including the deletion of entries from the system logs),

In the case of the most popular system software, i.e. MS Windows, there are two basic types of attacks: attacks on network protocols (e.g. SMB (Server Message Block), MSRPC (Microsoft Remote Procedure Call) and NetBios protocols) and the attacks on the implementations of web services (e.g. the non-standard implementations of standard protocols such as HTTP, SMTP and POP3).

The most frequently found vulnerabilities in Windows systems concern: install legal operating system, tool and protection software, web server and services, workstation service, remote access service, MS SQL server, web browser, authentication, file sharing applications, e-mail clients, IM communicators.

For instance, the workstation service in Windows systems that is responsible for the access of users to files and printers recognizes whether the requested resource is included in the local system or is made available remotely, and it correspondingly redirects the inquiry. This service includes a vulnerability which may be used through appropriate preparation of the DCE/RPC call since the problem consists in accepting the parameters transmitted to the login function without prior verification of their values. Windows system provides the local network users with access to files and printers with the use of SMB protocols or the Common Internet File System. The incorrect configuration of this service may lead in some cases even to gaining control of the system. Another threat is posed by the possibility of anonymous logging-in which makes use of the so-called null session (without giving the authentication parameters). These sessions may be used by the services run on the Local System account to communicate with other services on the network. For instance, in the case of Windows 2003 system, the access with the use of null sessions is allowed by the Active Directory service operating in the mixed mode.

Many problems concerning the Windows family platforms also result from the remote procedure calls (RPC). The RPC port mapper is published via the ports of TCP and UDP 135 and this service cannot be turned off since it influences the operation of the entire operating system. The MSRPC interfaces are also available on other TCP/UDP ports (139, 445, 593) and they may be also configured in such a way that they listen on the non-standard HTTP port through IIS (Internet Information Services) or COM Internet Services [2]. The identified vulnerability has not been removed even in Windows Server 2003 system in spite of introducing new security mechanisms. There are many malicious codes making use of this vulnerability (e.g. Blaster worm leading to DoS attack).

According to Microsoft, in the first half year of 2008, the number of vulnerabilities found in the computers of Windows system users fell by 4% in comparison with the second half year of the previous year and even by 19% in comparison with the period from January to June 2007. Unfortunately, the number of vulnerabilities with critical significance is increasing [3].

In the case of the Unix operating system, we can distinguish the attacks with remote access and the attacks with local access (the so-called attacks with privilege escalation). The remote access is gained through the network and the local access is gained with the use of the shell or through logging in the system. The attackers often begin with the remote use of vulnerabilities to gain access to the system as users and then they try to expand their privileges up to the root ones. The level of difficulty of these activities depends mainly on the type and configuration of a given operating system. The remote access to the Unix system can be gained through bypassing its security mechanisms with the use of one of four basic methods [2]: (1) attacking the listener services (to capture data); (2) routing through the Unix system; (3) infecting through the web page (when working on the network with the root rights); (4) attacking in the sniffing mode (making use of the vulnerabilities of the used sniffer).

The most frequently used vulnerabilities in the Unix systems concern: BIND server, web server, version control system, authentication, MTA (Mail Transfer Agent) server, SNMP (Simple Network Management Protocol), errors in configuration of NIS[1]/NFS, kernels. The most often the attacks concern the most popular software, hence the detectability of errors in its security systems is high. In the case of such operating systems as Linux or Mac OS, the number of malicious codes increases more slowly.

The causes of that can be sought in the fact that they are not much popular in China which is the country that is recently viewed as at the centre of development of malicious software.

25.3 The Vulnerability of Utility Applications to Attacks

The operating system is not the platform on which the attacks are carried out the most often. The applications are significantly more vulnerable layer. The attacks on the IT system are conducted with the use of errors in popular utility software: programs serving the electronic mail, text editors, spreadsheets, and web browsers. The use of all these tools is connected with the threat to the system security.

The web browsers are "in the front line". The most popular of them – Internet Explorer – is regarded as a tool which is very vulnerable to attacks. Pursuant to Security Focus Archive, 180 vulnerabilities have been discovered in Internet Explorer since April 2001. The vulnerability of the IT system to attacks is increased by the strong integration of this web browser with the kernel of Windows operating system (the information does not include Windows 7 and IE8). Of course, other known software of this type, like Mozilla Firefox, Opera, Google Chrome or Safari, also contains some vulnerabilities.

In March 2008, there was the information about the identification of vulnerabilities (including three critical ones) in the increasingly popular Firefox by Mozilla. The discovered errors cause - among other things – the instable work of the browser engine, in extreme cases - the interruption of its operation, allow for viewing the

---

[1] Network Information System – the potential source of information about the Unix network; the traditional attack consists in the use of the NIS tools to get the domain name which leads to gaining the NIS maps – dispersed critical information about the hosts in each domain.

history of visited web pages and for inserting the JavaScript code on another page. The next five security vulnerabilities (of which two critical ones) in Firefox were fixed with a patch by Mozilla in autumn 2008. The patches concerned, among other things, the errors in the scope of memory management and authorization granting. The discovered flaws allowed for introducing the malicious code through specially prepared web pages. In May 2009, there was the information about several errors in the Safari browser enabling the attacks on the operating system.

Pursuant to many experts, one of the most serious errors is the "zero day" vulnerability (the software flaw which has not been identified earlier and the way of its removal has not been described yet). Within a week of information about the vulnerability in Internet Explorer 7 browser, the hackers attacked 0.2% of IE users, which means that the problem concerned millions of people since the browser is very popular.

The browser producers strive for increasing the security of the tools and support the security systems of a computer in the subsequent releases of the software. Therefore, a new functionality of blocking the infected web pages has been introduced (Opera 9.5, Internet Explorer 8, Firefox 3).

Pursuant to F-Secure company, the favourite aim of the Internet hackers is popular Adobe Reader. From 1 January until the middle of April 2008, the PDF files were used by 128 dangerous attacks (the vulnerability allows the computer infection through opening the prepared PDF file), while in the same period of 2009 the company observed as many as 2305 such attacks. The alternative to Acrobat Reader is Foxit Reader; however, in May 2008, there was a warning against its serious vulnerability (FR 2.3 build 2825). A year after that there was the information that the malicious code called Gumblar which made use of vulnerabilities in PDF file readers infected thousands of web pages. These are only the selected examples of the vulnerability of utility applications to attacks. Recently it has been noted that the number of attacks on popular multimedia players such as Apple Quick Time, Real Player or Windows Media Player has significantly increased. For instance, in December 2008, the vulnerability was discovered in Realtek Media Player which allows for introducing and executing the malicious code (the error of buffer overflow occurs when opening the tracklist).

25.4 Hacking the Web Applications

The ways of data transfer have been changing along with the development of the web technologies. The integration of databases with the web servers and the use of dynamic methods of content generation caused that the portals and web pages are no longer only simple and static pages but also web applications. The growth of the user interaction with the web applications, in turn, increases the threat to the web resources. More and more often, the attacks on the IT systems are carried out not on the access level (protected by the Internet firewalls and system security mechanisms) but directly through the application layer. Also the HTTP used by the web applications to communicate with the user is quite vulnerable to attacks because it was created for static web pages.

The web pages are also more prone to different kinds of implementation and configuration errors due to their increasing complexity. The attacks on the web applications make use of the same methods as the attacks on servers. The search for weaknesses of applications involves analyzing their architecture and functionality, and other things. The specialized tools used for this purpose are Paros Proxy, SPIKE Proxy and Web Proxy. The attacks on the web applications are focused on such areas as authentication, session management, interaction with database, acceptance of input data.

The most popular methods of attacks include SQL injection, command injection, XSS (Cross-Site Scripting) and the attacks consisting in manipulation of the parameters sent between the user browser and the web application (manipulating the request chain, form fields, cookie values or HTTP request headers).

Many web applications make use of web server database query scripts in the popular SQL database language. In this case, the capture of queries sent by the script or their modification may lead to unpredictable operation of the application and result in unauthorized access to data or damage to the database.

The XSS attacks which have recently become popular consist in injecting the malicious code in a web page or a hyperlink and then infecting the computers of each visitor with it. The inserted software may include HTML code, scripts, ActiveX or Flash. The threat is also posed by the use of the Internet communicators and even of such protocols as SSH, SSL/TLS or IPSec which may be also attacked with the use of their flaws.

The example of searching for safer solutions for application servers includes new IIS releases (6.0, 7.0) where the web server's kernel has been separated from the code serving the application in such a way that they are separate processes. It causes that the applications work independently and the errors in the code of the user mode do not have any impact on the loss of application server security, which usually happened in the former releases. Yet, even if the new application releases have higher security level, the problem is posed by the widespread use of older solutions.

25.5 Application Vulnerabilities – The Frequency of Occurrence

Small and medium companies encounter the biggest number of problems with the corporate network security and the worst situation is with the wireless networks. Pursuant to the report by Napera company (from the end of 2008), three companies out of four do not control the security principles of their wireless networks. More than half of the research subjects could not even define which computers have access to the Internet and they did not check whether the software of such computers was updated on an on-going basis [3].

Secunia company published interesting results of the research carried out on 350 thousand users throughout the world. The test carried out with the use of "Software Inspector" program reveals that almost 30% of the used software does not have available system patches installed. The most dangerous applications include the multimedia players; the situation of browsers is a bit better (Fig. 25.1).

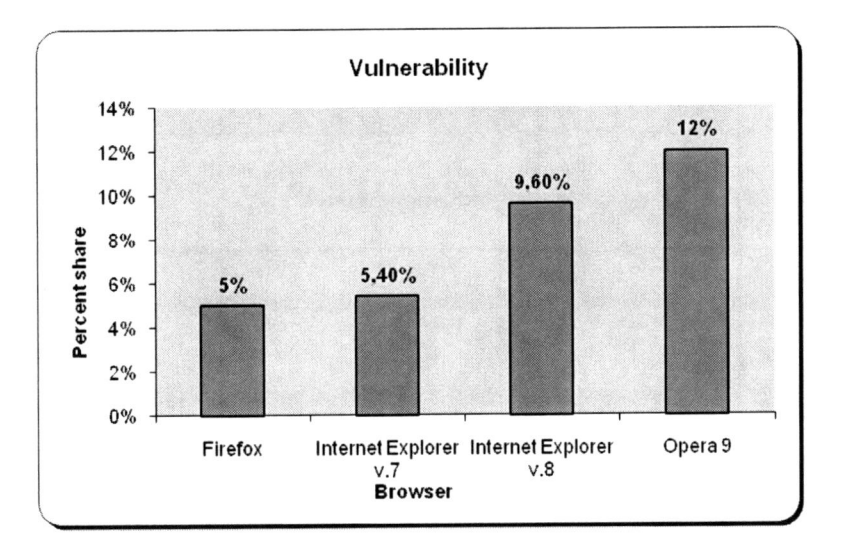

Fig. 25.1 The percent share of unprotected web browsers

The report published by Secunia at the end of 2008 shows that 98% of computers have at least one program which is not updated. The company gathered the data from 20 thousand computers whose users scanned their systems with the use of the free tool offered by this producer. Detailed results of the research are shown in Fig. 25.2; they are worse than the ones obtained a year earlier [3].

In autumn 2008, without giving any details, the public was informed about the discovery (by Grossman and Hansen) of exceptionally dangerous vulnerability in all web browsers. Pursuant to US-CERT, the problem is such that when clicking on a given website, a user may unknowingly click on the content of another website. The consequences of such attack may be extraordinarily dangerous and the possibilities of its conducing are unlimited. The representatives of Adobe, Mozilla and Microsoft have admitted that this is a tough problem with no easy solution.

In December 2008, the experts from Trend Micro company warned the Internet users against a new threat which had been disseminating with the use of a known vulnerability in all Windows system versions (including Windows Server). A worm

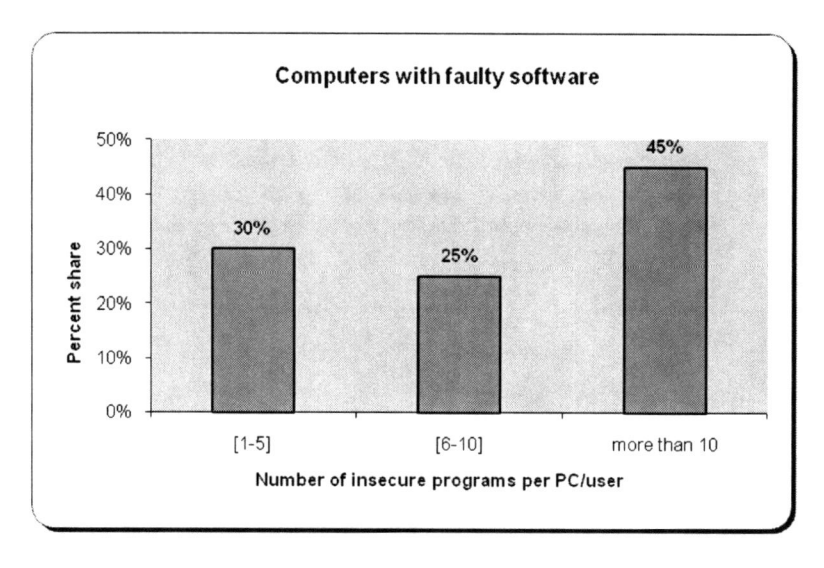

Fig. 25.2 The results of the system analysis in the scope of occurrence of software vulnerabilities

called DOWNAD.A is making use of MS08-067 patch to build the botnet. The most recent report suggests that it can also infect internal networks and the threat is of global nature [3].

25.6 Summary

The successful attacks on computer systems and networks are possible mainly as the result of the occurrence of numerous vulnerabilities in many popular operating systems, their network services and utility applications.

The known software flaws may be detected with the use of scanners and the solution to them is the on-going installation of system overlays or new corrected versions of a given program if they have been developed. Unfortunately, the time necessary to fix the vulnerabilities is not always short enough. The publishing of information about the threat prior to delivering appropriate patches is criticized because it happens that the fixes are released after months of waiting. In December 2008, The Inquirer website blamed Microsoft for the attacks on more than 10 thousand web pages within a week as a result of the tardiness of this company. The number of voices that fixing the software with patches is no longer effective is also increasing. During the conference organized by AusCERT in May 2008, the Security Director in Cisco, expressed the opinion that the use of patches is costly and ineffective since it is only a temporary solution to a problem. There was also a case when the publication of a patch was forgotten (Mozilla, December 2008).

References

1. Kurek, M., Lutynia, A.: Zagrożenia związane z udostępnianiem aplikacji w sieci Internet. In: Materiały konferencyjne i Ogólnopolska Konferencja Informatyki Śledczej, Katowice (2009)
2. McClure, S., Scambray, J., Kurtz, G.: Hacking zdemaskowany. Bezpieczeństwo sieci - sekrety i rozwiązania. PWN, Warszawa (2006)
3. Onet.pl, http://www.bezpieczenstwo.onet.pl
4. CanSecWest, http://www.cansecwest.com

Chapter 26
Security of Internet Transactions

Zygmunt Mazur, Hanna Mazur, and Teresa Mendyk-Krajewska

Abstract. State-of-the-art IT facilities are ever more frequently applied in all areas of life. Nowadays, Internet connection, online resources and services are taken for granted. It is much quicker (almost immediate) and more convenient to access electronic information and use online services than to employ traditional methods. Therefore, more and more people take advantage of these options. Aware of the benefits, they often forget about network threats (that arise from inappropriate software protection and failure to follow necessary security rules) or simply underestimate them. This chapter deals with threats to Internet transactions with particular emphasis on new ways of obtaining information and data under false pretences and threats to services rendered in wireless networks.

26.1 Introduction

Sustained development of electronic technologies completely changes our everyday life, the way we communicate, complete formalities, submit tax returns, do shopping, effect financial settlements, the sphere of science and entertainment, etc. The Internet is used to browse websites, read newspapers and books, watch TV, listen to the radio and communicate. More and more errands can be done via the Internet. Companies, shops and banks are extending their offer of non-cash settlements. Clients have more and more payment cards and e-accounts (and related passwords), and, at the same time, they conclude fewer cash transactions. It offers additional benefits as we can transfer money at any time of the day or night.

Zygmunt Mazur · Hanna Mazur · Teresa Mendyk-Krajewska
Wroclaw University of Technology,
Faculty of Computer Science and Management, Institute of Informatics
e-mail: {Zygmunt.Mazur,Hanna.Mazur,
Teresa.Mendyk-Krajewska}@pwr.wroc.pl
http://www.ii.pwr.wroc.pl

E. Tkacz and A. Kapczynski (Eds.): Internet – Technical Development and Appl., AISC 64, pp. 243–251.
springerlink.com   © Springer-Verlag Berlin Heidelberg 2009

New IT solutions make life much easier but also carry threats such as computer viruses, spam or network traffic eavesdropping aimed at stealing personal data. It is not possible to ensure 100% security of electronic transactions. Legal bank websites are often attacked and used to carry out attacks that allow for obtaining confidential data such as bank account passwords. The most cautious Internet users are the Americans and the French who are very mistrustful about making payments online. Research conducted by F-Secure among people aged 20-40 from different countries (the USA, Canada, Great Britain, France, Germany, Italy, India and Hong Kong) has shown that on average around 31% of people are afraid of making financial transactions via the Internet with a credit card. The most distrustful nations turned out to be the Americans and the French (more than 60% expressed their concerns) [4].

Computer espionage and intrusions have taken on such huge proportions that even the current President of the US, Barack Obama, joined the fight and on 29 May 2009 he announced the appointment of a special assistant (czar), within the White House, whose main duties will cover coordination of computer system protection against electronic espionage and terrorism. The number of attempted intrusions to Pentagon's computer systems is constantly increasing and in 2008 it amounted to 360m (6m more than two years earlier). Even the greatest armament program, Joint Strike Fighter (a multirole combat aircraft project), which is worth $300bn, has been hacked. In recent months, the cost of rebuilding the computer systems following the intrusions has reached $100m.

A report compiled by Kaspersky Lab reveals that in May 2009 as many as 42,520 unique pieces of malware and spam were detected on PCs. On the other hand, a report produced by Gemius shows that in 2008 66% of Polish Internet users did shopping on the Internet (Fig. 26.1), while 76% of the buyers did shopping at online auctions and 60% in Internet shops [5].

In 2008 Polish Internet users spent on online shopping over PLN 11bn, which means a 36.4 increase on the year 2007, at a 13% increase in traditional retail sales (Fig. 26.2) [1]. For comparison, in 2008 American Internet users spent on shopping $214bn (a 7% increase on 2007).

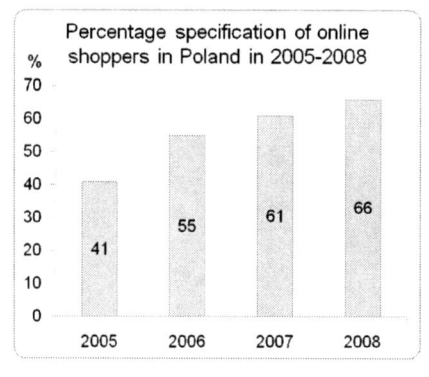
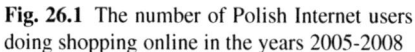
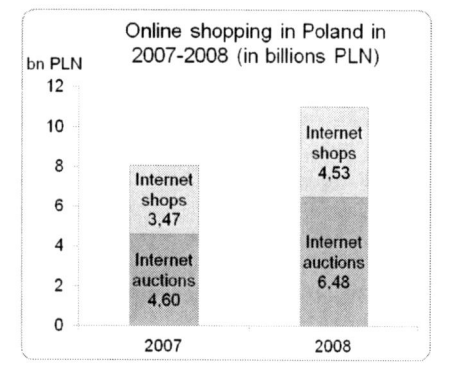

Fig. 26.1 The number of Polish Internet users doing shopping online in the years 2005-2008

Fig. 26.2 The value of e-commerce in Poland in 2007 and 2008

In Poland the Internet is used by 53% of men and 45% of women. The average age of Internet users is 35. The group of Internet users is made up mainly of better-educated people. The network is used by nearly all university graduates whereas persons with vocational training represent only 35%. The straight majority of network users are inhabitants of big cities – 65%, the smallest group of Internet users are people living in the countryside (35%) [2]. More and more people shop online – in January 2009 28% of adult Poles did shopping online (the year before the number was only 15%) and the most popular products were books, CDs, electronic equipment, clothes and footwear (chemical agents and medicines were less popular, whereas the least popular product was food).

Our decisions about buying products in a given Internet shop are affected by, among others: availability of the product and their price, trust in the company and in security of supplied data, opinion held about the company, previous experience with online shopping, user-friendliness of the application, available support.

26.2 Threats to Internet Users

In March 2007 McAfee published information concerning the degree of security of visited websites [3]. The report shows, among others, that: the most dangerous domains are .info and .com, the most secure domain is .gov, the most secure websites are the Scandinavian ones (Finnish, Norwegian, Swedish), the Polish domain .pl is 18th (out of 266 participating in the research) among exploit-threatened domains (programs that take advantage of errors in software in order to, for example, take control of the attacked system), 25th in terms of malware threat and 56th in terms of spam threat.

More and more often the media report different threats to WAN users. Even the protection software used may contain dangerous defects. In April 2008 Symantec informed about vulnerability in ActiveX control provided with Norton protection software (AntiVirus, Norton Internet Security, Norton System Works and Norton 360). The detected vulnerability may be used to take control of computers that use the Windows platform.

Recent commentaries posted on frequently visited websites have been spreading fake antivirus software called PrivaceCenter. Once we launch it, there is displayed a message informing about computer scanning, detection and removal of viruses, while malware is being installed in the background (e.g. a password capture software). So PrivaceCentre is a fake protection program installed by ZLOB Trojan horses. It imitates the Windows Vista Security Centre, generates false reports concerning the purported infection of the computer by malware and non-existent network attacks. Therefore, we need to be very cautious when using programs of unknown origin and when updating the system and protection programs. It often happens that attackers take advantage of soft errors quicker than producers manage to fix them (the so-called "zero day attacks", i.e. attacks targeted at new vulnerabilities for which no patches have been developed). Threats caused by such attacks are becoming ever more serious and the number of programs that take advantage of the 'zero day' vulnerability is growing (e.g. in October 2008 such programs accounted for 13% of malware whereas in November 2008 for as much as 30%).

In Poland, just like all over the world, online shopping is becoming more and more popular. According to data provided by the Poznań Supercomputing and Networking Centre (PSNC), in 2007 there were in Poland around 3 thousand portals offering goods on the Internet and the worth of goods purchased online amounted to PLN 8bn. While carrying out the transactions, customers are asked to provide personal data (full name, address, credit card No., etc.). A report published by the Security Team at the PSNC in January 2008 contains, among others, a description of incorrect mechanisms handling cookies that are used in the process of identification of users logged into an online shop [1]. Tests performed by the team (on fifty shops) revealed that 32% of the shops store session data in a catalogue/tmp on hosting companies' servers, which is not safe as it allows other users being logged in to download the data.

According to the OWASP standard, cookies that identify a logged-in user should be stored 5 minutes at the maximum while 2% of shops set their lifespan at 7 days, 4% - 1 hour and as much as 94% set the time at the end of the session [6]. It poses a serious threat in the case a session is abandoned without logging out as the system does not recognize closure of a web browser (and the lifespan of file-based sessions is 24 hours). Once the user logs out, 72% of shops close the session on the shop's server within 30 minutes whereas 8% – within an hour and 20% within 24 hours! As much as 70% of shops allowed users to assign numbers to their sessions. In 98% of cases session numbers were not connected with IP numbers, which means that the transactions were carried out via popular unencrypted wireless connections.

A report compiled by the Office Trading Fair shows that the British spend on online shopping twice as much as inhabitants of other European countries. In 2008 they made purchases worth £44bn and 54% of the British polled state that online shopping is safe (in 2006 only 26% of respondents hold such an opinion).

In March 2008 it was reported that a botnet created based on D-link routers to carry out DDoS attacks was found on the Internet. The infection spreads through 23/TCP and most likely the malware, which takes control of the router, takes advantage of a vulnerability in the Busy Box application. In the same month users were warned against PDF files circulating in the network (posted on websites or sent as email attachments) that took advantage of a vulnerability in Adobe Reader and Acrobat applications (in the fragment supporting the JavaScript code). The detected defect allows for buffer overflow and execution of any code in the system. Once the malicious file is opened, a Trojan horse (Zonebac) is downloaded, which, among others, disables anti-virus programs and modifies results of website retrieval.

Data security is seriously threatened during mass sports and entertainment or IT events such as, for example, CeBIT (Centrum der Büro- und Informationstechnik), where a lot of people use WLANs that lack adequate protection. Particular cautiousness should be exercised when using free hotspots which may upload data from notebooks while they are connected to the Internet.

The popular Google search engine also may contribute to theft of confidential information from poorly protected servers. As a consequence, in March 2009 19thousand of payment card numbers, owners' data, expiry dates and CVV codes fell into the wrong hands. The cards (issued by several different companies – Visa and MasterCard, to name a few) belonged mainly to the Americans and the British.

Data about websites visited by a user (URL and other data sent between the client and HTTP) can be accessed by, among others: network administrators, employers, Internet service providers, other users of a given computer (e.g. from the history created by a browser).

The persons may track websites visited in the network by a given user. Confidential data may be unintentionally lost for example when the data are sent by an employee to their private e-mail account (created in a free hosting service) in order to work at home.

Recently we have witnessed attempts of penetrating social networking services such as "nasza-klasa". Many users of such websites disclose detailed information about themselves and their friends this way providing data used to carry out dedicated attacks targeted at selected persons. In May 2009 there appeared a new worm which attacked thousands of users of Twitter. After clicking on a link sent in an email, people were redirected to a fake Twitter website where they were asked to provide their user name and password. If they keyed the requested data, malicious e-mails were sent to all persons who visited their profile while the user was redirected to a dating website. Within 2 days as many as 13000 Twitter users fell victims to the criminals. Many websites systematically inform about threats such as spam and viruses (for example, www.arcavir.jn.pl, www.viruslist.pl).

In June 2009 Kaspersky Lab employees identified a 25-million malware – a new modification of the Koobface worm called Net-Worm.Win32.Koobface, which attacks users of popular social networking websites (Facebook, MySpace). This is another instance showing that social networking websites are the target of ever more frequent attacks. Net-Worm.Win32.Koobface spreads in a very simple way: users of social networking websites receive a message allegedly sent by their friend, which contains a link to a video clip on an unknown website. When the users try to play the film, there appears a message encouraging them to update Flash Player. However, instead of an update a Koobface worm is installed. It functions as a backdoor, which allows it to carry out on the computer the instructions it receives from a remote administration server.

According to the Open Web Application Security Project (OWASP) that deals with the security of Internet applications, the major weaknesses of Internet applications are as follows: broken authentication and session management; vulnerabilities that enable Cross-Site Scripting (XSS), i.e. inserting, on a website being attacked, a code (usually in JavaScript) which can infect the computer of the user viewing the site; vulnerabilities that enable SQL-injections; susceptibility to Cross-Site Request Forgery (CSRF); malicious file execution; improper or no error handling; providing other entities with information about vulnerabilities, errors and technical details of applications - information leakage; insecure direct object reference; insecure cryptographic storage; insecure communications; failure to restrict URL access; lack of

control of input data (possible buffer overflow); defective access control and inadequate authority assignment; inadequate pseudorandom number generators; incorrect
algorithm implementation; no data encryption; improper storage of passwords, keys
and certificates; improper configuration.

26.2.1 Drive-by Download Attack

The use of web browsers (when looking for files with free software, music, films,
some phrases and others) is connected with the risk of running malware, mainly
due to poor computer protection. When Internet users use a browser, there are many
background processes going on they are unaware of, such as hidden download and
installation of malware from the network. A report compiled by ScanSafe reveals
that 74% of malware identified in Q3 2008 on Internet users' computers were downloaded from infected websites viewed by the users. In the last ten years the use of
e-mails for sending malware has decreased while websites and instant messengers
have become a very popular medium applied for that purpose.

Attacks on computer systems often take advantage of software defects. To this
end attackers apply exploits – malicious, easily accessible programs that can be
used even by those who have no experience in hacking. At present each security
scanner is equipped with an up-to-date base of popular software vulnerabilities.

Drive-by download attacks use a web browser that connects Internet users to
servers that contain exploits [6]. A malware is installed on a user's computer without their knowledge and permission. The attack is carried out in two stages. A user
visits a website that contains a malicious code which redirects the connection to an
infected server which belongs to a third person and contains exploits. The exploits
may take advantage of vulnerabilities of a web browser, unpatched plugins to the
browser, ActiveX formats susceptible to attacks and errors in other manufacturers'
software. Before an exploit is downloaded, a user may be an unlimited number of
times redirected to other websites. Moreover, legal websites are becoming ever more
popular target of hackers, who post there a redirecting code that launches attacks via
the browser. If an exploit is effective, it can install on the computer for example a
Trojan horse, which will allow the hacker to get unlimited access to the computer
being attacked, steal confidential data or carry out a DoS attack. In 2007, following hacking of the popular website of the Bank of India and numerous redirections,
Windows system users were redirected to a server with an email worm, two rootkits,
two Trojan horse downloaders and three backdoors. The attack on the Bank of India's website involved obfuscation with the use of JavaScript, numerous redirections
with the use of iFrame and fast-flux technology. In December 2008 there appeared
an exploit called "Curse of Silence", which circulated among cell phones (some
versions of Symbian S60) and blocked the function of texting. The exploit was very
easy to use and the relevant instructional video was available online. Many mobile
phone operators responded immediately and started filtering the texting traffic in

order to block messages containing the exploit. Nokia launched a free tool called "SMS Cleaner".

Most Internet browsers (Firefox, Internet Explorer, Opera) contain mechanisms that block malicious programs and warn users against entering infected websites. However, it does not ensure full protection. In order to avoid drive-by download attacks, users should scan the operating system and tool software, paying special attention to vulnerabilities, and systematically install patches and updates, use browsers that block infected websites, use a firewall, protection software and an Internet traffic scanner, avoid pirated software and software of unknown origin.

26.2.2 Attack through SSLStrip

In February 2009 in Washington, at a Black Hat conference devoted to IT security, a hacker presented his program used in Man-in-the-Middle-Proxy-type attacks called SSLStrip, which changes all HTTPS calls a user keyed into HTTP ones (as a general rule, when calling a website, users do not precede addresses with a string defining a secure protocol https:). It allows for stealing data that users use to log into services such as Yahoo!, Google, PayPal.

The SSLstrip attack can be carried out for instance in public Wi-Fi networks. It converts websites that are usually protected with the SSL (Secure Sockets Layer) protocol into unencrypted versions. The tool misleads both parties - the server and the user - keeping them convinced that encryption is active. Internet users tend not to pay attention to the fact whether an address begins with HTTP or HTTPS, therefore, they do not notice the swap (for a website not supported by SSL). During an experiment "attackers" managed to intercept 200 SSL connection requests, 114 Yahoo! And 50 Gmail passwords as well as data of 16 credit cards within 20 hours.

All state-of-the-art mobile phones with settings defined with the use of the OMA (Open Mobile Alliance) standards are susceptible to such attacks. Providers of smart phones claim that we can protect ourselves against such attacks in different ways, for example, a network operator can filter all texts containing setup data which do not come from the operator or block access to DNSs outside its IP addresses.

26.3 Eavesdropping in Computer Networks and Scanning of Ports

Information about software, configuration and activated services can be obtained by eavesdropping network traffic and scanning ports. The data gathered enable the attacker to identify weaknesses of the system that allow for effective attack on the selected target. Depending on the access to eavesdropped packages we can distinguish two types of eavesdropping: passive (which consists in reading the content of packages – used when an intruder has direct access to data being sent), active (when an intruder does not have access to the desired packages and redirection of network traffic is required).

In order to identify services being activated attackers scan ports since many services are by default connected with an assigned port number. Information about services being used on a given computer allows attackers to determine the type and version of the operating system, used applications and point out vulnerabilities in the security system.

Information about services prone to attacks can be obtained by intercepting banners (e.g. messages about opened sessions with the use of programs such as netcat or amap). Hackers try to apply only those tools that allow them to remain anonymous.

Wireless networks are particularly easy to penetrate, and such penetrations are difficult to detect. For the purpose of searching and documenting the WLAN, (standard 802.11) attackers can apply different tools with various functions, which allow them to: analyse headers of packages and initialization vector fields, verify SSIDs/ESSIDs (Service Set Identifier, Extended SSID), verify MAC (Media Access Control) addresses, identify applied security (WEP, WPA) or inform that no security is used, verify the signal range, obtain information about protocols used, obtain information about IP addresses, crack the encryption key.

Wireless networks are considered less secure than the cable ones. Their weakness is the medium – open and available to a prospective intruder. Research conducted by Motorola has shown that from among 400 companies in Western Europe (employing more than 1000 persons) as much as 65% apply the same security to protect wireless networks as in the case of cable networks while wireless connections require much better security.

26.4 Summary

Both e-commerce and e-banking are becoming more and more popular. E-commerce applications require ever more effective protection to ensure highest possible level of security. Clients' data are stolen in both cable and wireless networks, ATMs (Automated Teller Machine) and POSs (Point of Sale).

Internet users must be aware of the network-related threats. They have to be cautious even when they have protection software installed because there emerge new threats and malware, which are not detected by the installed protection tools straightaway. In order to effectively fight modern threats, Internet users must constantly broaden their knowledge and use comprehensive, integrated anti-virus solutions that ensure effective multi-layer protection.

References

1. Bezpieczeństwo sklepów internetowych. Sesje i ciasteczka. Poznańskie Centrum Superkomputerowo-Sieciowe, Poznań, security (2008),
   http://www.psnc.pl/reports/sklepy_internetowe_cookies.pdf

2. Public Opinion Research Center, `http://www.cbos.pl`
3. e-biznes.pl,
   `http://www.e-biznes.pl/inf/2007/20336,McAfee_Mapa_zlosliwej_sieci.php`
4. Gemius, `http://www.gemius.pl`
5. Global Security Week, `http://www.globalsecurityweek.com`
6. Viruslist.pl, `http://www.viruslist.pl/news.html?newsid=530`

Chapter 27
Security of Internet Transactions – Results of a Survey

Zygmunt Mazur and Hanna Mazur

Abstract. Security of Internet transaction is very important. It is not only the installed software, type of hardware used and the protection used by service providers (shops, banks, etc.) that influences the level of security; the behaviour of users (the parties of transactions, account holders, etc.) plays an importance role too. The overall level of security is influenced by many complementing actions. This chapter presents results of a survey conducted among students on the security of Internet transactions. Its aim was to assess students' actions in this field.

27.1 Introduction

Research conducted by the Public Opinion Research Centre shows that in Poland in January 2009 there were 16.7m Internet users aged 15-75. Nearly 50% of the Poles polled can access the Internet from home, 70% use the Internet to seek information they need to perform their work, and 70% of the respondents use instant messengers, mostly Gadu-Gadu (92% of instant messenger users) and Skype (67%). Email is used by 39%, whereas 20% of users have at least one account in social networking services [6].

In 2007, at the Global Security Week Conference, there were published "ten ways how to protect privacy"; we should, among others: have limited confidence in network co-users and resources; use encrypted connections, if possible; use a separate credit card (with a limited capital) when doing shopping online; control the history of transactions made, and definitely use a protection software (anti-virus, anti-spyware, etc.) that is systematically updated (including the virus base) [5].

Zygmunt Mazur · Hanna Mazur
Wroclaw University of Technology, Faculty of Computer Science and Management, Institute of Informatics
e-mail: {Zygmunt.Mazur,Hanna.Mazur}@pwr.wroc.pl
http://www.ii.pwr.wroc.pl

E. Tkacz and A. Kapczynski (Eds.): Internet – Technical Development and Appl., AISC 64, pp. 253–260.
springerlink.com

Online privacy is about the right to decide to whom, when and which data we want to disclose and a possibility to view the data, edit and delete them. The right to privacy was for the first time defined in 1890 by Samuel D. Warren and Luis D.Brandeis in Harvard Law Review as "the right to be let alone" [2]. In September 2007, at a UNESCO conference, Google Inc. called for establishing standards in Internet privacy protection. Today as much as around 75% of countries do not have any legal regulations protecting privacy. In the US the regulations differ depending on the state or sector, in European countries many legal regulations were adopted before the Internet boom (in the mid 90s) therefore they are outdated. A lot of information, including confidential and personal one, is available in countries in which it is not protected by any regulations.

Internet users must be aware that very often when visiting websites they provide a lot of information about themselves and their computers [4]. Programs installed for statistical purposes record, among others, the number, days and hours of visiting websites, IP address, domain, subdomain, city (voivodeship, country, continent), browser used, type of operating system, proxy, way of entering a given website (directly or indirectly and in the case of the latter - via which website). Furthermore, they may record words and phrases entered in the browser, complete data about recent references, etc. The acquired data are used, for example, in various analyses aimed at precise determination of users' preferences and interests. More and more websites offer also various free services (e.g. legal regulations and their interpretation, medical advice, archive issues of magazines, free versions of software, travel agencies' catalogues in a printable format) in exchange for personal data (e.g. an email address, phone no., mailing address, etc.). Before a user provides their personal data, they should check the company's privacy policy. It is important whether the company is recommended by a trusted institution (such as, for example, BBBOnline or Trust) that controls information protection. In order to protect privacy we can use programs that remove from the system records about our activity on the Internet (gathered in cookies[1] or in the form of history created by our browser).

A serious threat are the ever more popular social networking services and blogs, which allow for collecting various information, e.g. on the place of residence, property and valuables, number of co-habitants, lifestyle (work, trips) and social contacts, keeping a dog, etc. Blogs like Twitter are particularly dangerous as they are updated in real time and provide current (valid) information about a given person. Photos published on the Internet are another rich source of information, especially if they are systematically geotagged[2], as they allow for establishing the person's whereabouts and time of absence from home. Other popular tools are services used for localizing people such as bliziutko.pl, a Polish service by Gadu-Gadu Network. The service (and a location-aware application at the same time) is to facilitate meeting and integration of people living in the same place, taking joint actions, writing commentaries, collecting photos of the vicinity, etc. However, the data gathered and published in social networking services are very sensitive, therefore it

---

[1] Small portions of data sent by the Web server and recorded on the user's PC.

[2] Geotagging consists in assigning relevant place (e.g. names of places, geographic coordinates, height above sea level) and time stamps to data (photos, events, websites).

is recommended to exercise utmost caution when sharing personal data that often seem irrelevant. A similar service has been provided by Google. Google Latitude also systematically informs about our whereabouts and tracks our friends.

A report produced in 2007 by CERT Poland [3] reveals that there emerged a new threat to Internet users in the form of companies that operate on the edge of the law, which provide clients (guaranteeing anonymity) with their own lines and servers, both physical and virtual ones, so that they can post illegal content (connected with terrorism, Nazism, racism, etc.).

27.2 Electronic Bank Accounts

People associate the beginnings of e-banking in Poland (mBank, Inteligo) mainly with high interest on accounts and deposits as well as lack of other services such as loans. Despite the high risk, the Poles entrusted banks with their savings. At present, almost all banks, both Internet and traditional ones, offer Internet bank accounts supported by the network. Keeping such accounts reduces costs.

Internet accounts have numerous advantages: convenience, quick access to the account balance and history, a possibility to systematically track all financial operations, including the amount of charged fees and interests on the capital. Connecting the credit card with the bank account allows for controlling both the balance and the credit card debt. Such an account allows also for insuring a car or a house, joining a pension fund, opening a deposit account, to name a few.

Towards the end of 2008, the Polish Bank Association saw more than 6.4m active e-banking users in Poland, of which 653thousand represented the SME sector. In mid 2008, 28.1m clients had payment cards (of which 19.1m used debit cards) while 2.9 payment cards are equipped with microprocessors. In May 2009 more than half of adult Poles did not have a bank account. In order to stimulate non-cash transactions, the Polish Bank Association and the National Bank of Poland suggested sending all pays and pensions to bank accounts which would force people to open bank accounts.

With adequate precautions taken, e-banking guarantees high degree of security. Moreover, it is much less stressful to transfer money from a computer at home, irrespective of the amount, than to withdraw money at a bank counter and ride with the cash on public transport.

Online shopping is a type of electronic commerce used for business-to-business (B2B) and business-to-consumer (B2C) transactions over the Internet. The first pilot system was installing in 1984 in Tesco in the UK (first demonstrated in 1979 by Michael Aldrich). The first B2B solution was used by Thomson Holidays in 1981.

The risk that e-banking may entail is, among others: identity theft, that is an attempt to acquire confidential information (logins and passwords) by masquerading as a trustworthy entity; lack of physical evidence of conducted on-line transactions; lack of responsibility on the part of banks for consequences of revealing an ID or password to a third party, which allow for accessing an account.

In order to ensure security of online transactions, users must:

- install legal operating system, tool and protection software,
- avoid using private accounts on public computers (at work, university, Internet cafes or hot spots),
- follow recommendations and procedures concerning use of the account,
- have limited trust in received mail they have not expected,
- create a separate bank account with a limited amount of money,
- keep correspondence related to online shopping,
- key addresses of online banks and shops (instead of using links received in mails as they may redirect to fake websites).

27.3 A Survey 2008 on the Security of Internet Transactions

There was a survey in 2008 which assessed precautions taken by IT students of Wroclaw University of Technology while making online transactions [1]. As much as 77% of the students polled lost their data due to computer or software failure – Fig. 27.1(a), whereas 41% had problems arising from malware – Fig. 27.1(b).

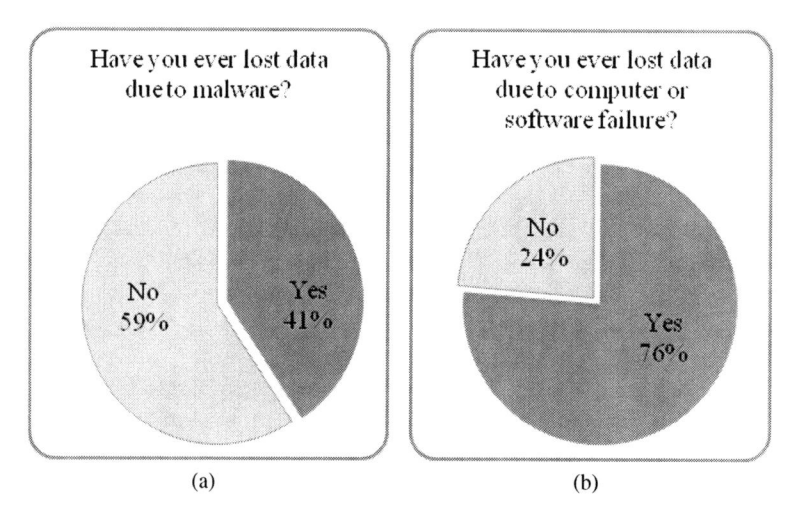

(a)   (b)

Fig. 27.1 Loss of data: a) due to malware, b) due to computer or software failure

Fig. 27.2 presents responses given on protection software installed by the respondents (on their PCs), of the following type: anti-virus, anti-spam, anti-spyware and firewall. The necessity to install such software seems obvious the more as students use different computers at home and at the university, and often transfer data between the two places. Therefore, protection of computers against malware is a must, also due to the importance of the projects (large and time-consuming) they prepare.

Meanwhile, the survey showed that not all the respondents take due precautions in this respect. Their behaviour is the more incomprehensible as the Internet offers numerous free computer protection programs.

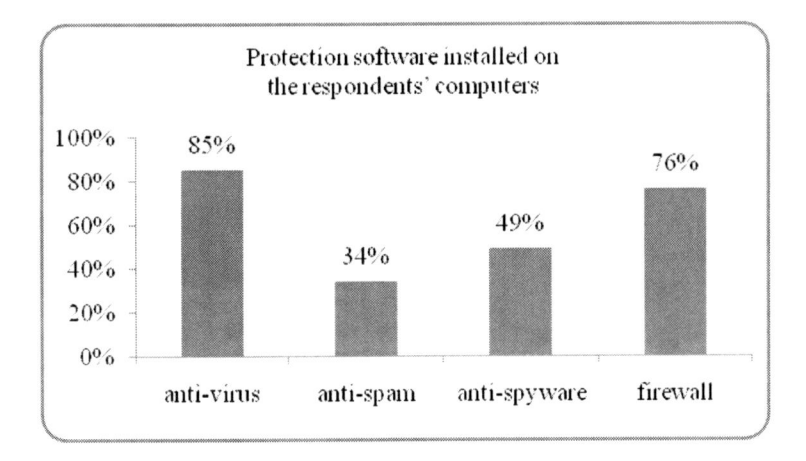

Fig. 27.2 Program protection used by the respondents

The use of different passwords to different accounts and the frequency with which they are modified is of great significance when conducting financial transactions online. The survey revealed that 83% of the respondents make financial transactions via the Internet (Fig. 27.3).

Fig. 27.3 A pie chart showing the use of the Internet for conducting financial transactions

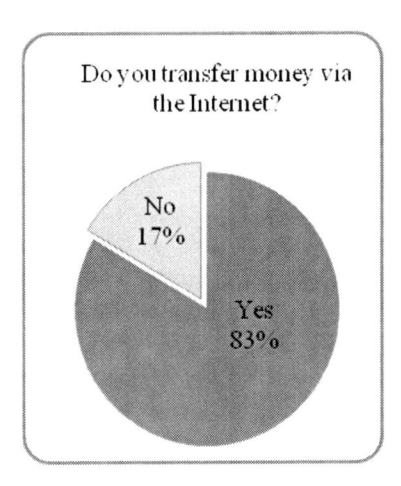

77% of the respondents use different passwords for different accounts (Fig. 27.4) while as much as 41% do not change their passwords at all.

The frequency of changing passwords is presented in Fig. 27.5.

Fig. 27.4 Assignment of different passwords to different accounts by the respondents

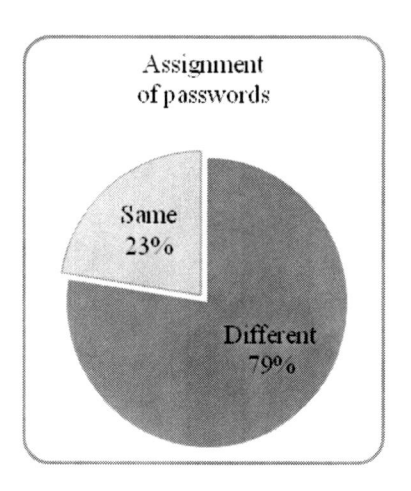

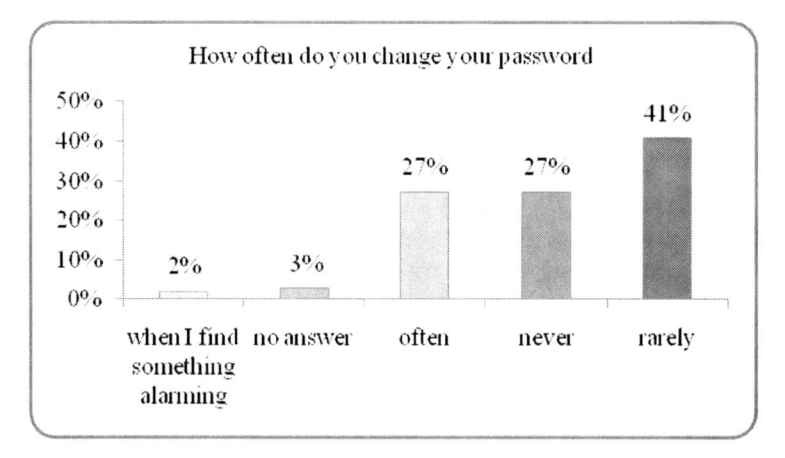

Fig. 27.5 The frequency of changing passwords by the respondents

The transactions usually concern small amounts (Fig. 27.6) but only 29% of the respondents have a separate bank account opened for that purpose, so they risk that in the case their account is hacked they may lose their savings.

From among 103 respondents, 37 persons use strong passwords, 57 persons both (strong and weak) and only 9 persons use weak passwords (Fig. 27.7).

The group of the students surveyed is familiar with the subject of security on account of the faculty of their studies, specificity of performed tasks and interests. The respondents are aware of the significance of protection of their computer resources and the necessity to make back-up copies every day, install protection software, and manage accounts and passwords correctly. Meanwhile, the results of the survey show that the basic rules of computer security are neglected although this is frequently the cause of losing data and even, as a result, failing to receive a credit for a course.

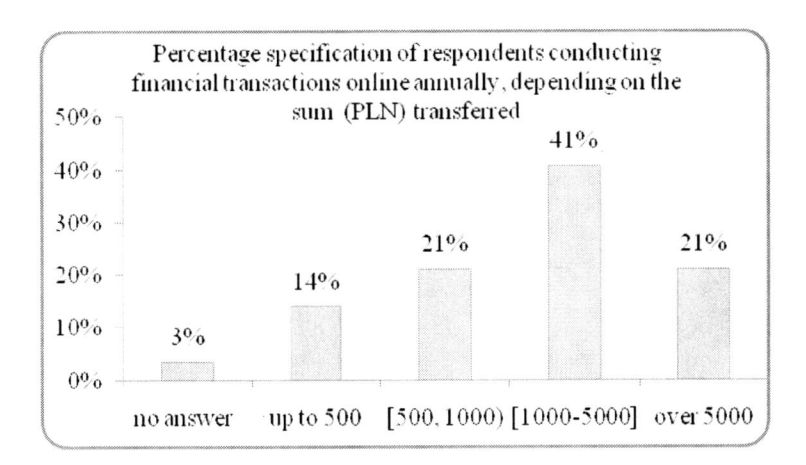

Fig. 27.6 Percentage specyfication of respondents conducting financial transactions online annually, depending on the sum transferred

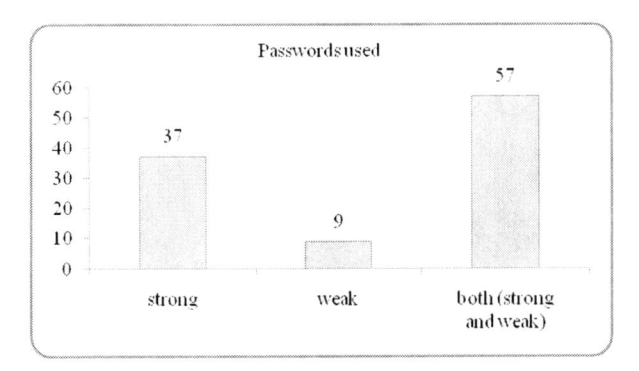

Fig. 27.7 Application of strong and weak passwords

27.4 Summary

Of great importance for the security of internet transactions is the propagation of knowledge about threats and their prevention among employees and IT staff, as well as end-users.

The weakest link in financial transactions conducted over the Internet is the customer who lacks proper training or acts carelessly. Results of the conducted survey support this thesis. Not all students take all necessary precautions even though they use the Internet to carry out transactions and have experienced computer malfunction and data loss.

References

1. Mazur, H., Mazur, Z.: Bezpieczeństwo komputerowe – zalecenia i praktyka. In: Huzar, Z., Mazur, Z. (eds.) Zagadnienia bezpieczeństwa w systemach informacyjnych, pp. 175–188. WKŁ, Warszawa (2008)
2. Warren, S.D., Brandeis, L.D.: The Right to Privacy. Harvard Law Review IV(5) (December 15, 1890)
3. CERT Polska, http://www.cert.pl/raporty
4. Gemius, http://www.gemius.pl
5. Global Security Week, http://www.globalsecurityweek.com
6. Public Opinion Research Center, http://www.cbos.pl

Chapter 28
Ontological Approach to the IT Security Development

Andrzej Białas

Abstract. The chapter presents researches on the ontology applications in the Common Criteria methodology. The first issue concerns the ontological representation of the standard security specifications, and the second one presents how this approach can be applied to elaborate the evidences for the IT security evaluation. Both issues are exemplified by the developed ontology and related knowledge base. Current research results are concluded and the planned works are shortly discussed.

28.1 Common Criteria Primer

The chapter deals with security engineering. The ISO/IEC 15408 Common Criteria standard (CC) [1] is the leading assurance methodology providing dependable IT solutions for applications when the assessed risk is high or the appreciated asset value is significant. The assurance is the confidence that the IT products or systems meet the security objectives specified for them, i.e. built-in security functions related to these objectives and representing measures will be effective when threats occur.

The assurance is measurable by evaluation assurance levels (in the range EAL1 to EAL7) and depends on the rigour applied to the development. The assurance level is claimed by developers who provide evidences that the given EAL is met. The IT product or system, called there the target of evaluation (TOE), its security specification, called the security target (ST) and the related evidences are evaluated allowing to get the certificate [2].

The CC methodology includes the following processes:

- the IT security development process, related to the elaboration of the security target (ST) for the given TOE, specifying finally the TOE security functions;

Andrzej Białas
Research and Development Centre EMAG, 40-189 Katowice, ul. Leopolda 31, Poland
e-mail: a.bialas@emag.pl

E. Tkacz and A. Kapczynski (Eds.): Internet – Technical Development and Appl., AISC 64, pp. 261–269.
springerlink.com

- the TOE development process with the use of the assumed technology;
- the IT security evaluation and certification process (not discussed there).

The main stages of the IT security development process [1] are: the security problem definition, security objectives, security requirements, and the TOE security functions work-out. During this process, the TOE security model is developed and refined. Developers need precise and commonly understandable specification means - design patterns, to fill in the model structure with the security interrelated contents. The CC standard provides security functional (SFR) and assurance (SAR) components as the specification means for security requirements only, but the specification means called generics are defined by developers freely for other stages. Some existing relations concern the given stage, e.g. expressing the mutual supportiveness of the specification items, and some relations deal with the items of neighbouring stages. The latter express how the security problem is solved by the security objectives, how these objectives are mapped to the SFRs, and how these SFRs items are implemented by the TOE security functions.

The second CC key process is the TOE development. It includes the elaboration of evidences that the TOE and its security functions meet the EAL requirements. The evidences concern: the TOE development, guidance documents, life-cycle definition, testing and vulnerability assessment issues. The evidences details are implied by the SARs components of the claimed EAL. The issue is how to compose these elementary requirements into the form of documents representing evidences provided for the evaluation. Developers need patterns and procedures allowing to elaborate evidences for the given TOE and the claimed EAL. During the refinement of the TOE security model to the TOE model, many complicated relations should be controlled as well.

28.2 Introduction

Both discussed processes require design patterns, like specification means and evidences templates, and the ability to be the master of their relationships across the models. These issues can be organized as a knowledge base, allowing to retrieve and reuse design data and facilitating the change management. To achieve this the author proposes to apply the ontological approach. The ontology represents formal specifications of the terms in the given domain and relations between them.

The ontologies allow to analyze, share and reuse knowledge in an explicit and mutually agreed manner [3]. Ontologies were elaborated recently in many disciplines including information security.

The most relevant works deal with the modelling of:

- the CC security functional requirements (SFRs); the SFRs are mapped to the earlier specified security objectives - both expressed by ontologies; the mapping is performed with the use of the GenOM CC ontology tool [4];
- the CC security assurance requirements (SARs); the ontology based tool [5] is used to support evaluators, not IT security developers.

There were other ontologies in information security, expressing: basic concepts, security algorithms, security functions, attacks and defence issues, and trust - [6], [7], the security of services, agents, assurance and credentials - [8], risk and security management issues [9], [10], [11], [12].

None of them:

- offers specification means for all stages of the CC-compliant IT security development process, i.e. the dedicated set of terms expressing threats, organizational security policies (OSPs), assumptions, objectives, SFRs, SARs and security functions according to the CC methodology,
- is able to express the entire IT security development process and the evidences elaborated for the evaluation process.

28.3 The Ontological Approach to the Common Criteria

The paper is based on the author's earlier works [13], [14] and the monograph [15] featuring the CC compliant IT security development framework (ITSDF). The UML/OCL approach was used to model data structures and subprocesses related to the IT security development process. The models of the specification means were elaborated too. The means include CC components and the introduced semiformal generics, called enhanced generics, as their features are similar to those of CC components, allowing parameterization, operation, derivation, iteration, etc. Finally, the computer tool supporting developers was built. The chapter presents continuation of these works based on new possibilities offered by the ontological approach.

The first results of these researches were presented in the paper [16] whose objective was to improve the ITSDF framework by introducing the Security Target Ontology (STO), encompassing the concepts related to all IT security development stages. The paper [17] presents selected issues concerning the Specification Means Ontology (SMO), the paper [18] - its validation on the firewall-, and the [19] - on the motion sensor project example. Both these ontologies were integrated into ITSDO (IT Security Development Ontology), expressing the whole CC methodology.

ITSDO has been developed on the basis of knowledge engineering principles [3] and with the use of the OWL (Web Ontology Language) and the Protege Ontology Editor and Knowledge Acquisition System from Stanford University [20].

The ITSDO ontology elaboration process is typical and begins with the domain and scope definition. It requires to investigate the ontology-related matters, i.e. different aspects of the IT security- and TOE- development processes. It allows to define the ontology terms within the domain and the ontology competency questions. The terms deal with both these processes. The competency questions are issues that the ontology-related knowledge base is able to answer. For ITSDO they can be: " What are the security objectives to enforce a given OSP or counter a given threat?", " How to uphold (by security objectives) a given assumption?" or " What are the predefined security functions concerning, for example, the biometric identification or trusted communication path?". Next, the hierarchy of classes and class properties are defined. The class hierarchy creates the taxonomy of terms in the discussed domain. Some classes have an abstract meaning, some have instances (individuals). The individuals represent specification means (generics, functional and assurance CC components, evidences and their patterns, etc). Knowledge acquisition, creation of individuals and filling in their properties (slots) allow to build a knowledge base. The ontology and knowledge base need permanent tests and validation on different designs.

For the IT security development process the ITSDO ontology provides design patterns related to the security target model and to the specification means (enhanced generics and CC components). The first example presents these facilities.

Example 1: The ontological approach to the security target specification.

The example deals with the motion sensor for a digital tachograph system [21]. The Fig. 28.1 presents the first group of design patterns, concerning CC specifications, i.e. the security target (ST), protection profile (PP), their low assurance versions, and their parts used to compose them. They are represented by a part of the ITSDO concepts hierarchy (ontology classes) shown on the left in the " Class Browser" window. Some classes have instances (individuals) whose numbers are placed in the brackets. The " Instance Browser" in the middle shows the list of the currently defined individuals of the highlighted SecurityTarget class. One of them is the ST_MotionSensor individual whose properties (slots) are shown in the " Individual Editor" window on the right. Parts of the ST [1] are expressed by object-type slots, filled in by the appropriate individuals. Please note others: the assigned-ToProject slot used to manage of the project elements, hasEAL slot expressing claimed EAL, hasSetOfEvidences slot pointing to the evidences, and hasProduct-Category slot showing the classification category - the same as used on the Common

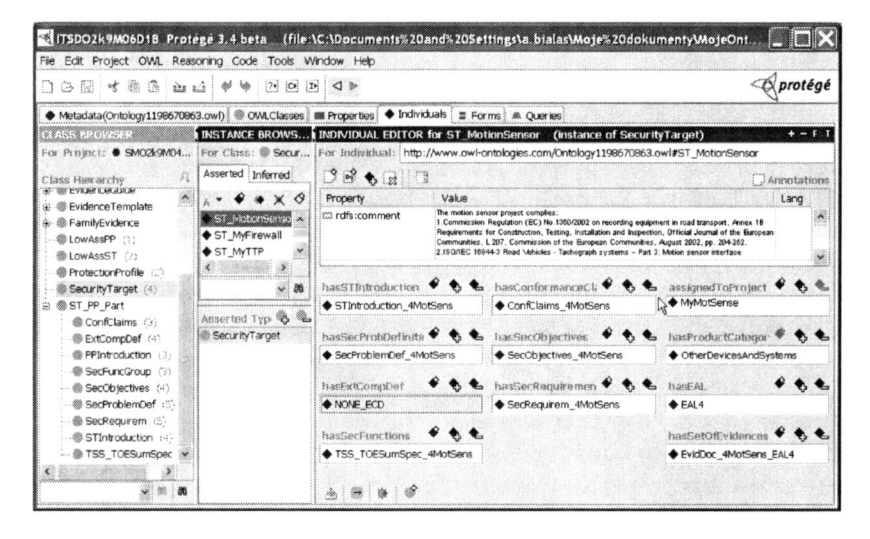

Fig. 28.1 The ontological model of the motion sensor ST, shown in the Protege tool [20]

Criteria portal [2]. The annotation type slot, rdfs:comment can be used to document any ontology element.

The Fig. 28.2 exemplifies the second group of design patterns, i.e. the specification means (author's defined enhanced generics and CC-defined components) used to fill in the CC specification structures with proper contents. The "Class Browser" shows the part of the ITSDO class hierarchy of the enhanced generics (a taxonomy). For example, the TDA_Generic class represents direct attacks against the TOE, the TIT_Generic class - attacks against the TOE IT environment, while TPH_Generic - attacks against the TOE physical environment. For others, please refer to the [15], [17]. The "Individual Editor" presents slots of the SecProblemDef_4MotSens individual. There are three main elements: threats, OSPs and assumptions. Moreover, the threat agents, legal subjects and protected assets are also expressed by the enhanced generics which can be used as parameters of other generics. A generics name contains the prefix representing the generic family, e.g. "TDA_", followed by the "mnemonics" expressing the semantics of the given generic, e.g. TDA_Access concerns illegal access to the motion sensor [21]. Inserted generics can be refined on the project level by adding extra comments changing their semantics. Clicking on the given generic, its details (expressed by slots contents as well) are displayed in a separate window (not shown). For the threat type generic this can be: description, details of a derived version, if it exists, threatened asset, threat agent, exploited vulnerabilities, related asset value, likelihood, risk value, and first and most the security objectives being able to counter this threat. The last property represents the example of the relation between the elementary security problem and its solution. Moreover the ITSDO ontology provides design patterns for the TOE development process in the form of templates for evidences, relationships and procedures, showing how to use these templates, based mostly on the [22] guide. The elementary evidences

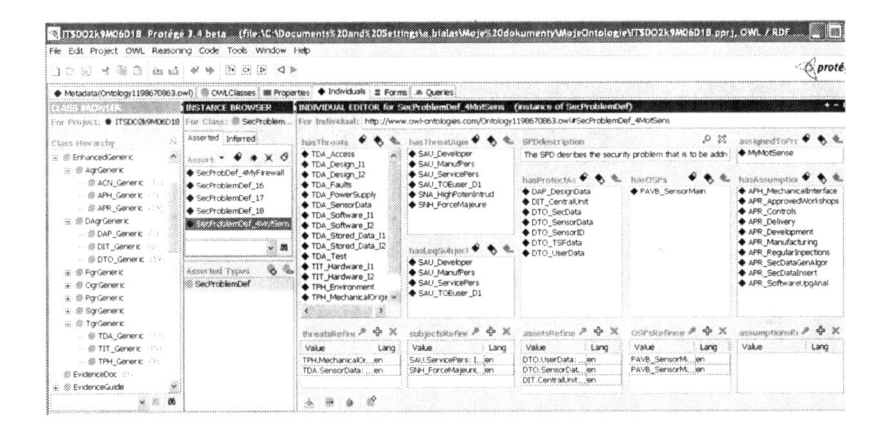

Fig. 28.2 The ontological model of the security problem definition of the motion sensor [20]

implied by each SAR are used as the building material to compose the set of evidences for the given EAL and TOE.

Example 2: The ontological representation of the evaluation evidences.

The second example concerns evidences elaborated for the MyFirewall project based on the [15]. The set of evidences (refer to the slot hasSetOfEvidences in the Fig. 28.1 is represented by the EvidDoc_4MyFirewall_EAL4plus individual, shown on the left part of the Fig. 28.3. Apart from some textual slots, it contains three sources of elementary evidences used to compose the set of evidences. The hasBasicEvidence slot includes elementary evidences implied by the SARs from the EAL package, hasEvidFromSubstSARs - implied by SARs replaced by others of the higher rigour, while hasEvidFromAddedSARs - implied by the optionally added SARs [1]/part 3. The right part of the Fig. 28.3 shows the elementary evidence example, i.e. ADV_FSP_EAL_4. It concerns the ADV_FSP.4 Basic functional specification SAR component, dealing with the TOE interfaces and included in the EAL4 package [1]/part 3. The ADV_FSP_EAL_4 is elaborated on the Tmpl_ADV_FSP_4 template basis, using guidelines (procedure) pointed by Guide_ADV_FSP_fam (guides concern particular assurance families, while templates particular components).

Currently a simple Protege built-in query mechanism is used to retrieve any issues dealing with the CC-methodology ontological representation.

Example 3: The simple knowledge base query for the MyFirewall example.

The Fig. 28.4 shows retrieving the TOE security functions and evidences. The details for the highlighted SFDP_FwlAdminAuth function, providing the authentication of the firewall administrator, are shown in the separate window. This generic implements the following functional components: FIA_UID.2 User identification before any action, FIA_UAU.2 User authentication before any action and FIA_AFL_1 Authentication failure handling [1]/part 2.

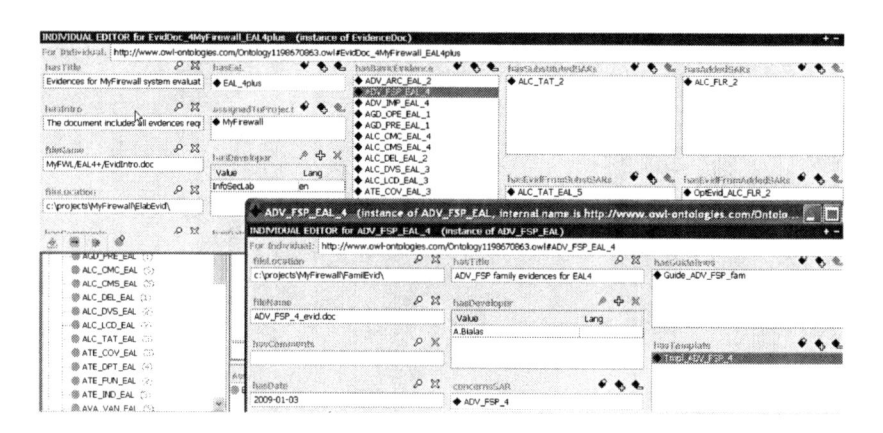

Fig. 28.3 The MyFirewall evidences management using the Protege tool [20]

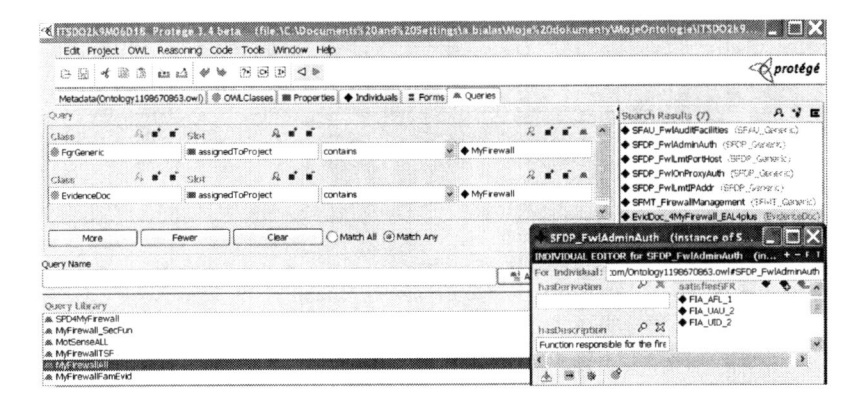

Fig. 28.4 Using the Protege query [20] for the MyFirewall project

28.4 Conclusions

The achieved results shows that the whole Common Criteria development methodology can be represented as an ontology and the related knowledge base. The main advantages of the applied approach are the improved design preciseness and reusability. The performed validations on the nearly real projects show that this approach helps:

- to improve the common understanding of the used terms by IT security developers, evaluators, users and other stakeholders (CC consumers),
- to express models relationships (horizontal, vertical) in the security projects,
- to transform the unified and abstract CC model to the concrete IT product or system model during its development with the use of the given technology.

Containing hundreds knowledge base items, ITSDO is still a prototype which requires: knowledge base extension, further development, introducing more sophisticated competency questions, and more validation on real projects.

Acknowledgments

This work was conducted using the Prot?g? resource, which is supported by the grant LM007885 from the United States National Library of Medicine.

References

1. ISO/IEC 15408. Common Criteria for IT security evaluation. Part 1-3
2. Common Criteria portal, http://www.commoncriteriaportal.org/

3. Noy, N.F., McGuiness, D.L.: Ontology Development 101: A Guide to Creating Your First Ontology, Knowledge Systems Laboratory (2001),
   http://www-ksl.stanford.edu/people/dlm/papers/
   ontology-tutorial-noy-mcguinness-abstract.html
4. Yavagal, D.S., Lee, S.W., Ahn, G.-J., Gandhi, R.A.: Common Criteria Requirements Modeling and its Uses for Quality of Information Assurance (QoIA). In: Proc. of the 43rd Annual ACM Southeast Conference (ACMSE 2005), vol. 2, pp. 130–135 (2005)
5. Ekelhart, A., Fenz, S., Goluch, G., Weippl, E.: Ontological Mapping of Common Criteria's Security Assurance Requirements. In: Venter, H., Eloff, M., Labuschagne, L., Eloff, J., von Solms, R. (eds.) New Approaches for Security, Privacy and Trust in Complex Environments. IFIP, vol. 232, pp. 85–95. Springer, Boston (2007)
6. Vorobiev, A., Bekmamedova, N.: An Ontological Approach Applied to Information Security and Trust. In: 18th Australasian Conf. on Information Systems, Toowoomba (2007),
   http://www.acis2007.usq.edu.au/assets/papers/144.pdf
7. Almut Herzog's web site "Security Ontology", Linkoping University,
   http://www.ida.liu.se/~iislab/projects/secont/
8. Kim, A., Luo, J., Kang, M.: Security Ontology for Annotating Resources, Naval Research Laboratory, Washington (2005),
   http://chacs.nrl.navy.mil/publications/CHACS/2005/
   2005kim-NRLOntologyFinal.pdf
9. Elahi, G., Yu, E.: A Goal Oriented Approach for Modeling and Analyzing Security Trade-Offs. In: Parent, C., Schewe, K.-D., Storey, V.C., Thalheim, B. (eds.) ER 2007. LNCS, vol. 4801, pp. 375–390. Springer, Heidelberg (2007)
10. Ekelhart, A., Fenz, S., Goluch, G., Riedel, B., Klemen, M., Weippl, E.: Information Security Fortification by Ontological Mapping of the ISO/IEC 27001 Standard. In: Proc. of the 13th Pacific Rim Int. Symp. on Dependable Computing, Washington DC, USA, pp. 381–388. IEEE Computer Society, Los Alamitos (2007),
    http://publik.tuwien.ac.at/files/pub-inf_4689.pdf
11. Ekelhart, A., Fenz, S., Klemen, M., Weippl, E.: Security Ontologies: Improving Quantitative Risk Analysis. In: Proceedings of the 40th Hawaii International Conference on System Sciences, Big Island, Hawaii, pp. 156–162. IEEE Computer Society Press, Los Alamitos (2007)
12. Tsoumas, B., Dritsas, S., Gritzalis, D.: An Ontology-Based Approach to Information Systems Security Management. In: Gorodetsky, V., Kotenko, I., Skormin, V.A. (eds.) MMM-ACNS 2005. LNCS, vol. 3685, pp. 151–164. Springer, Heidelberg (2005)
13. Bialas, A.: Semiformal framework for ICT security development. In: The 8th Int. Common Criteria Conference, Rome, September 25-27 (2007),
    http://www.8iccc.com/index.php
14. Bialas, A.: Semiformal Approach to the IT Security Development. In: Zamojski, W., Mazurkiewicz, J., Sugier, J., Walkowiak, T. (eds.) Proc. of the Int. Conf. on Dependability of Comp. Sys. DepCoS-RELCOMEX 2007, pp. 3–11. IEEE Computer Society, Los Alamitos (2007)
15. Bialas, A.: Semiformal Common Criteria Compliant IT Security Development Framework. Studia Informatica 29 2B(77) (2008), http://www.znsi.aei.polsl.pl/
16. Bialas, A.: Ontology-based Approach to the Common Criteria Compliant IT Security Development. In: Arabnia, H., Aissi, S., Bedworth, M. (eds.) Proc. of the 2008 International Conf. on Security and Management (WORLDCOMP 2008), Las Vegas, pp. 586–592. CSREA Press (2008) ISBN#1-60132-085-X
17. Bialas, A.: Ontology-based Security Problem Definition and Solution for the Common Criteria Compliant Development Process. In: Proc. of the Int. Conf. on Dependability of Comp. Sys. DepCoS-RELCOMEX 2009, pp. 3–10. IEEE Computer Society, Los Alamitos (2009)
18. Bialas, A.: Validation of the Specification Means Ontology on the Simple Firewall Case. In: Proc. of the Int. Conf. on Security and Management, WORLDCOMP 2009, Las Vegas. CSREA Press (July 2009) (accepted)

19. Bialas, A.: Security-related design patterns for intelligent sensors requiring measurable assurance. Przeglad Elektrotechniczny (Electrical Review), 92–99 (2009); R.85 NR 7/2009, ISSN 0033-2097
20. Protege Ontology Editor and Knowledge Acquisition System, Stanford University, `http://protege.stanford.edu/`
21. Commission Regulation (EC) No.1360/2002 on recording equipment in road transport, Annex 1B Requirements for Construction, Testing, Installation and Inspection. Official Journal of the EC, L 207, pp. 204–252 (2002)
22. Guidelines for Developer Documentation according to Common Criteria version 3.1, Bundesamt fur Sicherheit in der Informationstechnik (2007)

Chapter 29
Enhancement of Model of Electronic Voting System

Adrian Kapczyński and Marcin Sobota

Abstract. This article presents issues related to electronic voting, which can be used in elections to the parliament, presidential elections or national referendums. The advantages and disadvantages of such voting and protocols implementing the voting via internet are discussed. On the basis of Kutylowski model of e-voting an enhancement related to assigning rights phase is introduced.

29.1 Introduction

In everyday life we rely increasingly on the use of the computer and computer related technologies. Rather than use of the services offered in the traditional manner, one can use the electronic equivalent by application of the computer and the Internet which saves time and money. Instead of visiting a bank branch (the cost of the journey, the time lost to get there and possibly in the queue) in order to carry out the money transfer, one can log in to online bank account and within a few seconds perform such a transfer. You can use these services for 24 hours a day, which is unquestionably very convenient.

Some actions require the use of the computer and the Internet. For example, the only way to start the procedure of become a student at the Silesian University of Technology is to go through the process of recruitment, which is conducted by an electronic system called "SOREK". To participate in the recruitment process, each candidate must sign up for the individual system account and another way just does not exist. It shows that the whole process starts with the electronic registration, which can be done only with a computer connected to the Internet.

These two examples are the cases when the use of computer is the result of choice or necessity. However, there a number of examples where traditional solutions are

Adrian Kapczyński · Marcin Sobota
Silesian University of Technology,
Faculty of Organization and Management, Roosevelt 26-28, 41-800 Zabrze, Poland
e-mail: adrian.kapczynski@polsl.pl, marcin.sobota@polsl.pl

E. Tkacz and A. Kapczynski (Eds.): Internet – Technical Development and Appl., AISC 64, pp. 271–277.
springerlink.com   © Springer-Verlag Berlin Heidelberg 2009

the only ones which are applicable. An example of this is general elections in Poland - citizen wishing to vote, must visit the local election commission. It is not possible to do it via the Internet. This restriction will affect the level of electoral turnout, where the barrier before the voting could be a factor as prosaic as the rainy weather. It is, therefore fully reasonable to consider constructing a system which utilizes modern technologies - build e-voting system. Such a system is composed of two models: physical (infrastructure) and logical (protocols). Both models combined shall be the base for implementation of experimental e-voting system and tested in out-of-laboratory conditions.

The goal of the article is to introduce a new model of voting via Internet basing on the development of KutylowskiŠs model of e-voting. The models are presented just after short introduction to electronic voting: idea and protocols.

29.2 Electronic Voting

Electronic elections are not an emerging domain. There are many countries in which these elections have already been carried out, including USA, United Kingdom, Switzerland and Estonia. Everywhere, however, there were encountered numerous problems that have brought that to achieve a satisfactory level of choices made by the Internet far more expensive.

Key issues that emerged are:

- psychological barrier - the traditional vote is seen as a great event as the elections over the Internet will undoubtedly lead to the electoral, trivial action. In addition, e-elections require public confidence in the technical means and tools, which can be mixed. In the elections of 2007 in Estonia, a desire to vote by Internet has been more than 80 percent of people entitled to vote and put this way, only slightly more than 5 percent of the vote,
- technical problems - in the elections held in 2002 in the U.S. there have been instances of lost or incorrectly recorded votes. The scale of this phenomenon was significant - in some counties the number of such votes reached even 48 percent.

An Irish electoral system which was purchased for an amount more than 50 million EUR, was rejected by the Central Electoral Commission as a result of an independent security audit.

An additional barrier might be the law and legislation. In Poland, according to the law, the information system and ICT can only be means of supporting the work of the National Electoral Commission. It means that only voting cards and election protocols are applicable. In order to count the votes sent via electronic way we need to change the electoral law.

Examples cited here only mentions a number of problems associated with electronic voting. These are not, however, the key aspect of this article - more important are voting protocols which are discussed next.

29.3 E-Voting Protocols

To be able to carry out the electronic election (e.g. via the Internet) an appropriate protocol is required. The requirements of such protocol consist of the following [1]:

1. Only authorized persons can cast the votes,
2. Each authorized person is entitled to cast only one vote,
3. No one can determine for whom authorized person voted,
4. No one can duplicate other vote,
5. No one can change other vote,
6. All voters can check whether their votes were calculated.

There are a number of simple protocols, which could be used, but which do not meet all these conditions.

Protocol A [1] is a first example:

1. Each voter encrypts his vote with his private key,
2. In addition, it encrypts vote using the public key of Central Election Commission (CEC),
3. Each voter shall send their vote to CEC,
4. CEC decipher the votes, verifies the signatures, calculate the votes and announce results.

Protocol satisfies five of the six conditions, only condition number 3 is not fulfilled. Since the digital signatures of voters are used the CEC would be technically able to check who the person voted. This restriction does not apply to the Protocol B which is described below [2]:

1. Each eligible voter receives registration number from the Permission Central Agency (PCA). In addition, PCA stores all issued numbers in order the check that the person does not attempt to vote for a second time,
2. PCA sends the registration numbers to CEC,
3. a voter encrypts its vote and a registration number using public key of CEC,
4. CEC deciphers the package using the private key. Then check whether the registration number is on the list received from PCA. If the number is not present on the list, the votes shall be rejected and in other case the voice is counted, and the registration number is removed from the list,
5. after receiving all the votes CEC announces the results of the election and publish a list of registration numbers and the names of chosen candidates.

This protocol meets all established requirements. The only problem is that the PCA may contact the CEC in order to link the registration number with a name of person who has received that number, and later it could be checked, for whom the given person voted. There are similar protocols in which the PCA and CEC is one institution. Then identification of such person becomes even easier. Nevertheless, it seems that the minutes of such structures are a reasonable compromise between complexity and security.

If the only issue is the cover the personal data of a person who receives a registration number, than one can apply the protocol C:

1. Each entitled person shall be reported to CEC to obtain a registration number. CEC issues such a number and records it in order to verify later. The number is not assigned to a person, while the personal data of a voter that has received a number, are removed from the list of people entitled to vote,
2. Voter casting its vote completes the vote with received the registration number and uses CEC public key to encrypt the message,
3. CEC deciphers the message, check whether the vote contained in the registration number is entitled to voting and based on that, the vote is accepted or rejected,
4. After the end of voting, CEC announces results of the elections, together with a list of registration numbers and votes assigned with them.

On the one hand such an approach will result in that the CEC has only registration numbers, which will entitle the holders to vote without the possibility of their assignment to the particular person, and on the other hand, voters may (after the announcement of election results) check whether their vote is correct.

29.4 Model of e-Voting via Internet

The group of researchers under the direction of prof. Miroslaw Kutylowski (Wroclaw University of Technology, Poland) proposed the following model of assigning rights via Internet:

To vote online, you need to register in a particular state office by the legislature. They will also have two cards waiting for you: one for coding and one for the election. Each will hold a separate office: the election - Electoral Commission, while the coding - other institution, which we call the Mediator. At the time of registration you will receive the election Ű you shall choose one of the many cards, enclosed in an envelope. In addition, one of the intermediaries receive the coding. You can download any number of card encoding, but will receive only one card election. The intermediary is an institution, which prepares the card coding. The system can run multiple brokers, so that we can have greater confidence in the outcome of the election. Mediator may be not only the government but also a political party or NGO. You choose the mediator, the services which you use.

Detailed information about the conduct of the elections can be found at: www.e-glosowanie.org.

From the verification point of view presented model seems to be correct. It has only one weak point, namely, it is less clear how the cards are assigned to vote. It has been said about the selection of one of the many cards, but do not know what's next. Is the registration number of the card introduced into the system? If so, there is a vulnerability of the attribution of this number to the citizen verified. If the identification number of the card is not placed on the system at the time of collecting cards, but at the time of generation card to vote, what happens to the cards to vote, which had not been received by eligible? Whether their numbers are removed from the database of numbers that remain in the database? How CEC accounts missing cards ? It seems that in the whole model of voting over the Internet proposed by Kutylowski et al., the weakest point is in assigning e-voting right. It is not entirely clear whether this form of assigning rights on the one hand, will enable only people entitled to vote, and on the other hand, will not allow CEC to identify those persons with card numbers to vote.

29.5 Enhancement of e-Voting Model

Undoubtedly, the problem is the practical application of the point 1 of protocol C. It's easy to assume that the voter is identified as a person entitled to vote and then it is assigned an registration number to enable subsequent verification of the validity of the given voice. But how to assign a national identification number in such a way as CEC will not be able to assign this number to a specific person?

The proposed solution consists of the following steps:

1. Identification numbers are generated by a generator belonging to the CEC,
2. CEC has its local branches,
3. Citizens wishing to take part in the vote via the Internet go to LEC (Local Election Commission) appropriate for the place of residence. They may do so within the time limits laid down before the election,
4. LEC have lists of people entitled to vote,
5. Review the person entitled shall be made by a classical, not the digital way. The citizen shall sign the list of persons entitled to vote (as is the case with today's election in Poland) which results in the possibility of receiving an ID number that allows later cast the vote electronically. Such a person will not be able to vote in the traditional manner,
6. LEC after confirming the identity of the person entitled to vote electronically sends a request to the central database in order to generate an ID number,
7. sent ID number is printed at LEC and put in secure (darken) envelope and given to the voter. At the same time, the generated number is stored in the database identifiers in CEC, so on the election day to be compared with the ID number given by the person casting vote.
8. the next steps are the same as in protocol 3 starting from a point 2.

The novel idea is depicted on (fig.29.1).

What's the significance of such approach? When verification of the person entitled to vote is done in a classical way, and generating an identification number is done by electronic means from a central database it is protected against the possibility of linking these activities. Query posted by LEC to Central Generator means only that given person entitled to vote has reported to given LEC and that this person has been positively verified. It is not known, however, who it was, because there

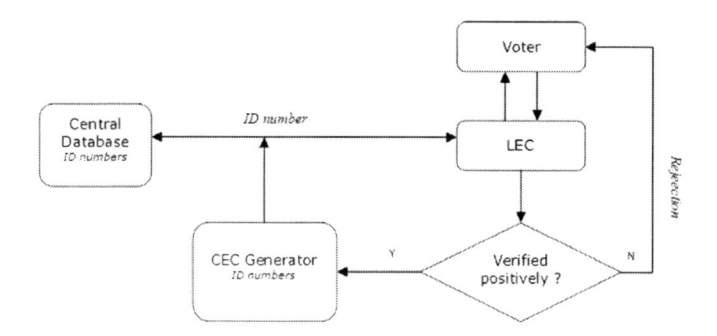

Fig. 29.1 Enhanced model of e-voting (assigning rights phase)

is no digital trace after the event. ID number sent by the CEC is invisible to LEC, so it is not possible to link a number obtained from a person with person who is receiving it. Before the election, the new lists are generated for the votes casted by the classical designated polling stations. These lists contain only the names of those who have not received identification numbers voting via the Internet.

29.6 Conclusions

No doubt e-voting will become a normal, ordinary action in the near future. It is very important therefore to provide long enough test period for fixing any problems with which may occur. Analysis of protocols of e-voting gives clear conclusion that the most difficult thing is to maintain two interrelated conditions: giving the permission to vote only to the entitled persons (identify the voter) and the identification of the vote in such a way that voters can recognize his vote, but not able to do so by someone else. It seems that the protocol proposed by the authors provides the fulfillment of these conditions while maintaining the criteria for assigning an identification number.

Further research work will be focused on delivering experimental e-voting systems utilizing proposed model (in logical layer), as well as the achievement of quantum cryptography and biometrics.

References

1. Schneier, B.: Kryptografia dla praktyków. Wyd. 2. WNT, Warszawa (2002)
2. Salomea, A.: Public-Key Cryptography. Springer, Heidelberg (1990)
3. Cohen, J.D., Fischer, M.H.: A robust and verifiable cryptographically secure election scheme. In: Proceedings of the 26th Annual IEEE Symposium on the Foundation of Computer Science, pp. s.372–s.382 (1985)

4. Cohen, J.D.: Improving Privacy in Cryptographic Elections. Yale University Computer Science Department Technical Report YALEU/DCS/TR-454 (1986)
5. Benaloh, J.C., Yung, M.: Distributing the Power of a government to enhance the privacy of voters. In: Proceedings of the 5th ACM Symposium on the principles in distributed computing, pp. s.52–s.62 (1986)
6. Benaloh, J.C.: Verifiable Secret-Ballot Elections. Ph.D. dissertation, Yale University, YALE/DCS/TR-561 (1987)

Author Index

LaVergne, TN USA
10 December 2009
166474LV00002B/12/P

9 783642 050183